Doubt-Free Uncertainty In Measurement

Colin Ratcliffe · Bridget Ratcliffe

Doubt-Free Uncertainty In Measurement

An Introduction for Engineers and Students

Colin Ratcliffe
United States Naval Academy
Annapolis
Maryland
USA

Bridget Ratcliffe
Johns Hopkins University
Baltimore
Maryland
USA

ISBN 978-3-319-12062-1   ISBN 978-3-319-12063-8 (eBook)
DOI 10.1007/978-3-319-12063-8

Library of Congress Control Number: 2014955354

Springer Cham Heidelberg New York Dordrecht London
© Springer International Publishing Switzerland 2015
This work is subject to copyright. All rights are reserved by the Publisher, whether the whole or part of the material is concerned, specifically the rights of translation, reprinting, reuse of illustrations, recitation, broadcasting, reproduction on microfilms or in any other physical way, and transmission or information storage and retrieval, electronic adaptation, computer software, or by similar or dissimilar methodology now known or hereafter developed.
The use of general descriptive names, registered names, trademarks, service marks, etc. in this publication does not imply, even in the absence of a specific statement, that such names are exempt from the relevant protective laws and regulations and therefore free for general use.
The publisher, the authors and the editors are safe to assume that the advice and information in this book are believed to be true and accurate at the date of publication. Neither the publisher nor the authors or the editors give a warranty, express or implied, with respect to the material contained herein or for any errors or omissions that may have been made.

Printed on acid-free paper

Springer is part of Springer Science+Business Media (www.springer.com)

Preface

Every time anyone takes a measurement, the only certainty is that the measurement is not exact. This is even true if you are a highly skilled and qualified metrologist; all measurements have some amount of uncertainty associated with them. For example, if we state that the temperature in a room is 70 °F, we do not mean that the temperature is *exactly* 70 °F. We probably mean it is close to 70 °F. Also, we do not mean that the temperature *everywhere* in the room is 70 °F. There may be hot spots close to radiators or running electrical equipment and cold spots near air conditioning vents.

Even "accurate" measurements have some degree of uncertainty. Take, as an example, the problem of weighing yourself on a bathroom scale. You may step on the scale once, or 10 times, or 100, or 1000. Let us assume that every time you step on the scale you record almost exactly the same weight. You may come to the conclusion that since all of the measurements were very close, you know your weight "accurately." But it is most likely that the bathroom scale has an error (for example, a zero offset) and you have very accurately got the wrong answer!

The concept of uncertainty is that there is always some doubt about a measurement. We can never remove all doubt, but we can aim to estimate and understand it. The theory and examples presented in this booklet aim to develop a quantifiable range within which the true value falls. This range, or uncertainty interval, cannot be determined exactly since we never know exactly how accurately any one measurement has been made. Consequently, when we quote uncertainty we always include a probability. This is the probability that the true quantity is within the quoted range.

So why is there an uncertainty in our measurements? Some uncertainty is due to the accuracy of the transducers we use to take measurements, and some is due to variability in the item being measured. In this booklet we will identify the most common sources of uncertainty. We will then see how these uncertainties can be combined so that we can identify the uncertainty in an actual measurement. Once we know the uncertainties in individual measurements, we will combine their uncertainties so that we can estimate the uncertainty associated with more complex engineering measurements.

The most recent theories of measurement uncertainty were first published by ISO[1]/BIPM[2] in 1992 in the "Guide to Uncertainty in Measurement," often referred to as the GUM. While the GUM is not a standard in the strict sense, it is the basis on which standards dealing with metrological subjects are drawn. Prior to the GUM, uncertainty analysis of measurements was an uncertain exercise requiring much good (or bad!) engineering judgment. Uncertainties and errors were categorized as either B-Bias or P-Precision errors, and the mathematical analysis often gave final estimates of uncertainty that were unrealistic and not useful. Since publication of the GUM, measurement uncertainties are grouped into two different categories: Type A and Type B. Type A uncertainties are those which are evaluated by statistical methods, and Type B are those which are evaluated by any other means. The new methods presented in the GUM give much more realistic and useful estimates of the uncertainty associated with individual measurements and calculated quantities.

Companies wishing to demonstrate their quality management system often get ISO 9001 certification. ISO 9001 is one of the most widely used management tools in the world today, and the most recent certification, ISO 9001:2008, has been in effect since November 2010. The next version of the standard is anticipated in 2015. Companies seeking ISO 9001 certification are required (among other things) to document transducer calibration and traceability. While ISO 9001 does *not* require all measurement systems to be calibrated, it does require calibration of those that contribute significantly to the control or verification of the quality of the product. One way companies can show whether a particular measurement system contributes significantly or not to their process is to perform an uncertainty analysis using the techniques outlined in the GUM and in this monograph.

This monograph explains the theory and methods in the GUM in a way that is approachable to the practicing engineer. We then present an extended case study. The case study presents many practical aspects of an uncertainty analysis in a form that can act as a template for the study of a different system. The case study also discusses different types of uncertainty budgets that can be developed using the results from an uncertainty analysis. Uncertainty budgets are a popular way of identifying where cost savings in a process can be made, or identifying whether a proposed change is cost-effective or not.

By necessity this monograph uses some statistics and calculus. We assume that most readers have seen this level of mathematics previously, but have probably forgotten the details! Therefore, we aim to explain the mathematics as non-mathematically as possible. The focus of this monograph is the understanding and application of uncertainty, and while we identify the essential mathematics we did not feel this booklet is the right place for detailed mathematics and statistics theory. If you want a book on mathematics or statistics, there are many excellent texts available on the market.

We hope that you find this book useful as you develop uncertainty analyses and budgets for your own systems.

---

[1] International Organization for Standardization.

[2] International Bureau of Weights and Measures.

Contents

1 **Terminology** 1
   Measurand 1
   True Value 1
   True Value and Uncertainty Interval 2
   Confidence, Significance and Coverage Factor (k-factor) 2
   Type A and Type B Uncertainties 3
   Systematic and Random Uncertainties 3
   Is it an Error or an Uncertainty? 4
   Elemental Uncertainties 5
   Calibration 5
   Measured or Calculated Quantities? 6
   Propagation of Uncertainty 7
   Standard Uncertainty and Expanded Uncertainty 7
   Root Sum of the Squares (RSS) 7
   The Big Picture 8

2 **Type A and Type B Elemental Uncertainties** 9
   Random Uncertainty 9
   Repeatable, or Systematic, Uncertainty 10
   Type A and Type B Elemental Uncertainty 11
   Sources of Elemental Uncertainty 13
   Accuracy—Pandora's Box? 17
   Final reminder 18

3 **Standard Uncertainty of a Measurement** 19
   Finding the Standard Uncertainty when an Uncertainty Is Quoted
   at a Certain Level of Confidence 20
   Finding the Standard Uncertainty when an Elemental Uncertainty
   Is Given as a Standard Deviation with No Information About
   the Sample Size 22
   Finding the Standard Uncertainty when the Uncertainty Is Given
   as a Standard Deviation and the Sample Size Is also Given 22
   Finding the Standard Uncertainty when the Elemental Uncertainty
   has a Known Statistical Distribution 25

Combining Several Standard Uncertainties to find the Standard
Uncertainty for a Single Measured Quantity 29

**4 Expanded Uncertainty of a Measurement and an Uncertainty
Budget for a Single Measurement** 33
Expanded Uncertainty, U 34
Example 35
Changing the Level of Confidence 35
An Uncertainty Budget 36

5 Propagation of Uncertainty & An Uncertainty Budget Example 39
General Principles 41
Absolute and Relative Uncertainty 43
Uncertainty Budget—How to Use the Uncertainty Analysis
to Improve the Accuracy of a Measurement Process 46
Uncertainty Budget Example: Radiation Heat Transfer 47
Conclusions 51
Earlier Examples Revisited 51

6 Sensitivity by Perturbation 55
Returning to the CNC Cooling Fluid Problem 57

Case Study 61
A Fully Worked Example Developing the Uncertainty Analysis
for a Process, Including a Discussion of Uncertainty Budgets 61
    The Big Picture 62
    Description of the Experiment to Measure Viscosity 63
    Measurements 65
    Uncertainty Analysis 65
    Transducer Uncertainty 66
    Summary of Transducer Standard Uncertainties 70
    Uncertainty of Measurement 70
    Propagation of Uncertainty 77
    Uncertainty Budgets 79
    Uncertainty Budget Summary 85

Appendix 87
The Mathematics of Resolution and Truncation 87
    Derivation of the Standard Uncertainty When Using
    the Full-Resolution 89
    Derivation of the Standard Uncertainty When Using
    the Half-Resolution 90
    Derivation of the Standard Uncertainty When Values
    Are Truncated 91
    Truncation with Error Adjustment 93
    The Standard Uncertainty for an Analog-to-Digital Converter 94

Terminology

Before we can delve into the qualms of uncertainty analysis, let us identify some critical terms and concepts that we will use later. We suggest that you first skim through this chapter to get an idea of terminology and content, and then come back to this chapter regularly as you work through the rest of the monograph.

In the real world different terms have different specific meanings, depending upon your specific branch of engineering and science. Therefore the descriptions below are exactly that—they are *descriptions,* and not legalistic definitions.

Measurand

The term 'measurand' is a generic one that refers to a quantity that is being measured. For example, if you are measuring the volume of gas pumped at the gas station, the measurand is the volume of gas, and if you are measuring wind speed, the measurand is the speed.

True Value

If we are weighing an object, then the 'True Value' is the actual value of that measurand (the weight). Sadly, we do not know the true values of any measurands, which is the reason we measure them. For example, if you are checking the electrical resistance of a resistor, the only reason you are measuring the resistance is because you do not know it. If you knew *exactly* the electrical resistance you would not have to measure it, and there would be no need for this monograph!

Since we do not know the true value, we take measurements. But each measurement has some uncertainty associated with it and the measurement is therefore only an estimate of the true value. The better the measuring equipment and the more trained the operator, the better the measured estimate is to the true value, but we can *never know the true value.*

> Situation: One technician quickly measures the size of an object and states that is it 1.34 in. A second technician undertakes an extensive series of measurements and states that the size is 1.3387±0.0002 in.
> Question: What is the true value of the size? I.e., what is the actual size?
> Answer: Even though it is *probably* between 1.3385 and 1.3389 in, we do not know the size *exactly*.

True Value and Uncertainty Interval

One way we can overcome the problem of never knowing the true value is to identify our best estimate of the true value, and quote an UNCERTAINTY INTERVAL, or range, in which we reasonably expect the true value to lie.

We now face the dichotomy of uncertainty analysis: In order to accurately identify the uncertainty interval in which a true value lies, it is necessary to know the true value. However, if we know the true value there is no uncertainty, the uncertainty interval is zero—and there is no need to take measurements! The consequence is that all uncertainty intervals are only *estimates* and therefore we always quote them with an associated probability that the true value lies within the quoted range.

Confidence, Significance and Coverage Factor (k-factor)

Uncertainty is often quoted at a certain level of confidence. For the previous length example, one way we could quote the results is to say that *"the true value of length is 1.3387±0.0002 in at 95% confidence."* This means that we are 95% confident that the true value lies within the quoted range.

An alternative way of giving the result would be to say that *"the true value of length is 1.3387±0.0002 in at 5% significance."* We are saying that there is a 5% chance of a *significant event*, the event being that the true value is outside the quoted range. Confidence and significance are related by:

$$\text{Confidence} + \text{Significance} = 100\%$$

A third way of giving the result is to quote a coverage factor, or *k*-factor: *"the true value of length is 1.3387±0.0002 in with a k-factor of 2."* The *k*-factor can be thought of as a multiplier that gives probability to an uncertainty. Mathematically the coverage (*k*-factor) is the number of standard deviations from the mean that include the percentage confidence. Coverage factors are most often based upon the normal distribution and later in this monograph we give a variety of *k*-factors and the probability they represent. For example, a *k*-factor of 2 represents 95% probability.

Type A and Type B Uncertainties

As we progress though the uncertainty analysis we will find that elemental uncertainties (see below) are assigned as either Type A or Type B. The distinction, as defined by the GUM, is that Type A uncertainties are those that are evaluated by the statistical analysis of a series of observations, whereas Type B uncertainties are those evaluated by any other means.

Systematic and Random Uncertainties

As a simplified example, let us return to the problem of weighing yourself using a bathroom scale. We will assume that the scale has not been zeroed correctly and consequently always under-reads by 5 lb. This is a systematic (or repeatable) uncertainty. Let us also assume that when you take repeated measurements they are all within 1 lb of each other. With a systematic uncertainty of 5 lb and a scatter of 1 lb, we can expect the readings will always be between about 4 and 6 lb too light. The following figure demonstrates the relationship between systematic and random uncertainties in a measurement. If the scatter caused by the random uncertainties is small compared to the systematic uncertainty, it is quite possible that the scatter of measurements will never include the true value, as is the case for the bathroom scale in this example, and the scale will never measure the correct true value.

If the systematic uncertainty is small compared with the random scatter, then the true value might lie within the random scatter, and the scale might sometimes measure the correct true value. Of course, you have no way of knowing if this is the case!

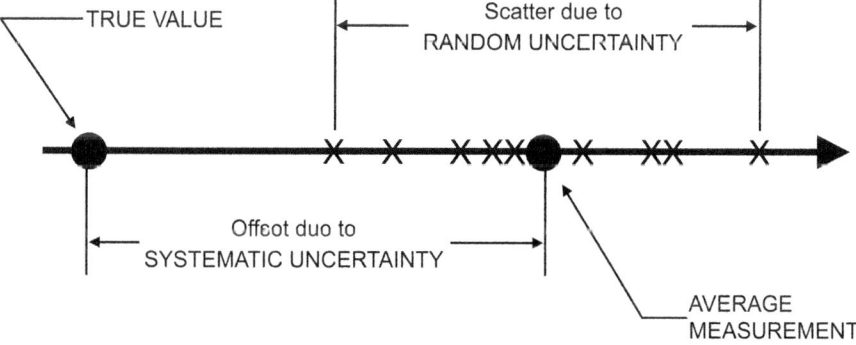

The terms systematic and random are "old" terms and as stated in the GUM, 'There is not always a simple correspondence between the classification into categories A or B and the previously used classification into "random" and "systematic" uncertainties. The term "systematic uncertainty" can be misleading and should be avoided.' Despite this declaration, the GUM continues to use these terms, and even includes the outdated terms 'bias' and 'precision'. Therefore, while recognizing the terms systematic and random are archaic, we will also continue to use them since they are descriptive of the types of uncertainty encountered in daily measurements.

One of the main aims of an uncertainty analysis is to estimate an uncertainty range (or interval) that includes the true value. It thus combines all types of uncertainty, including the systematic and random uncertainties.

> Scatter, systematic uncertainty and measurement uncertainty are not the same. Systematic uncertainty is a measure of how far the average measurement is from the true value
> Scatter is a measure of the random variability in a measurement.
> Measurement uncertainty includes both the scatter and systematic uncertainty.

Example:
The distance between two surveying targets is calculated from theodolite work station measurements of ranges and angles. Repeat measurements determine the distance between the targets is 427.75 ft. with a spread of ±0.03 ft. in the readings.

The next day the measurements are repeated, but using a different theodolite. These readings gave a distance of 427.85 ft. with a spread of ±0.02 ft.

Day 1: 427.75 with a spread of ±0.03 ft.
Day 2: 427.85 with a spread of ±0.02 ft.

Questions that might be asked:

Is one of the theodolites out of calibration?
Has one of the targets been moved?
What is the actual distance (i.e., the true value)?

Answer:
The spread of results is due to the random uncertainties only and does not include the systematic uncertainties. A full uncertainty analysis could well identify that the total uncertainty of this particular measurement is ±0.15 ft. Thus even though the two theodolites are yielding different distances, they could both be correct!

Is it an Error or an Uncertainty?

In this monograph we predominantly consider uncertainties. Uncertainty is a range of values in which we think the true value might lie and is usually associated with a plus/minus sign. For example, we might state that with 95% confidence a pressure is 75 psi ±2 psi, by which we mean that we are 95% sure that the true pressure is somewhere between 73 and 77 psi.

In comparison, an *error* is usually a residual amount that has been determined from a calibration. The error only has a plus *or* minus sign, and is an amount that can be calculated out of a measurement. For example, consider lab standard gage blocks that can be used to calibrate calipers. One particular gage block should be 2 in long, but calibration of the block showed that it is actually 1.99992 in long. The gage block has an *error* of −0.00008 in. This can be taken into account when the block is used to calibrate calipers.

> Uncertainties can never be known exactly. They must be estimated and quoted with a statistical degree of confidence.
> Errors are fixed and known, typically from a calibration. They can be taken into account, and thus do not actually affect the final measurement.

Elemental Uncertainties

Elemental uncertainties are those associated with a transducer or the parameter being measured. For example, suppose that you are using a surveyor's steel tape measure with 1/100 in graduations. One of the elemental uncertainties for the measure is related to the interval between the graduations (i.e., the scale resolution). The actual length of the steel tape itself will depend upon the ambient temperature—it will be longer in hot weather, and shorter in cold conditions. Thus, raw measurements will seem shorter in hot weather than in cold weather. For example, a steel surveyor's tape measure that is 100 ft. long will be about 0.075 in longer (more than 7 graduations) per 10°F increase in temperature. While some of this variability can be compensated for as an *error*, there is still some residual variation in measurement due to uncertainty in the temperature. This variation is another example of an elemental uncertainty.\

Calibration

This monograph does not investigate the intricacies of calibration, which is a full subject in its own right. Suffice it to say that often the combined elemental uncertainty of a transducer will be determined by a calibration laboratory, with the uncertainty being included on the calibration certificate. A transducer, or device under test (DUT), is calibrated by comparing it against a standard with a smaller amount of uncertainty. Industries commonly use a lab standard that has one-fourth of the uncertainty of the DUT (i.e., it is four times 'more accurate'). But how does one know the uncertainty of the standard? The answer is to compare it with another (higher grade) standard that has one-fourth of the uncertainty of the lab standard, and so on! Calibration is expensive and time consuming, and therefore it is common practice to use calibration services to calibrate and provide a calibration certificate for each transducer. The calibration company then becomes responsible for the traceability of the standards, as required by ISO.

One of the aims of calibration is to minimize or quantify the uncertainty in a transducer. While some systematic uncertainties (for example, those due to zero offset or sensitivity) can be minimized, they can never completely eliminated, and there will always be some residual uncertainty that we cannot measure or identify. After all, if we could identify a small zero offset, we could correct for it with a calibration error!

But be cautioned: just because, for example, the calibration certificate of a thermocouple states it has an uncertainty of $\pm 2°C$ it does not mean that temperatures measured with it have that accuracy. Each measurand also varies. For example, the actual temperature inside an autoclave may vary from place to place and time to time by $40\,°C$ due to variations of the process. This is an elemental uncertainty since it is associated with a measurand.

> The uncertainty in a measurement is a consequence of elemental uncertainties.
>
> Elemental uncertainties have two sources: Those due to the uncertainty of measurement in the transducer, and those associated with variability in the measurand.
>
> A calibration certificate should include an uncertainty which combines all the elemental uncertainties associated with the transducer itself.

Measured or Calculated Quantities?

Imagine that for your job you need to prepare a quotation for resurfacing a parking lot. To help with your quotation you might need to know the lot's surface area, and loosely you may say that you 'measured' the area. In actuality, it is more likely that you took a series of measurements (distances and angles) and then *calculated* the area—by hand or maybe using a total station that does the math for you. In uncertainty analysis we have to clearly differentiate between quantities that are actually measured, and those that are determined (calculated) from one or more measurements.

Measurands are measured *directly* with a *transducer* or *measurement device* and are called MEASURED QUANTITIES. Anything that is not directly measured, but is determined from one or more measured quantities is called a CALCULATED QUANTITY. For example, you may use calipers to *measure* the diameter of a shaft and then *calculate* its radius from the diameter. Or you may use a spherometer to *measure* the sagitta of a lens, and use the measured sagitta to *calculate* the lens' spherical radius. We will soon see that the distinction between measured and calculated quantities is important to uncertainty analysis. We will be able to determine an uncertainty in each measured quantity by referring to such things as a transducer's calibration certificate and the inherent variability of the quantities being measured.

However, in order to determine the uncertainty in a calculated quantity (such as the area of the parking lot) we will need to *propagate the uncertainties*.

Propagation of Uncertainty

Very often we cannot measure the thing we want. For example it is not possible to directly measure the volume of a piece of paper, and in cases like this we have to take several measurements and use them to calculate the quantity of interest. The different measurements may use the same transducer (for example, using the same ruler to measure the length and width of the paper) and sometimes different transducers will be used. For example, you may use a micrometer to measure the paper's thickness. You then calculate the volume as the product of the different measurements.

While we can estimate the uncertainty in each measurement from its own individual elemental uncertainties, this does not directly help us determine the uncertainty in the calculated quantity. For example, if you know the length of the paper with an uncertainty of ± 0.05 in (at 95% confidence), how do you use that value to work out the uncertainty in volume of the paper? After all, the length uncertainty is in units of inches, and the volume uncertainty will be in inches cubed. They have different units so the uncertainties must be different!

The procedure for determining the uncertainty in a calculated quantity based on the individual uncertainties for each measurement is called "uncertainty propagation."

Standard Uncertainty and Expanded Uncertainty

Standard uncertainties are given the symbol lower-case u. Expanded uncertainties are given the symbol upper-case U.

We can think of the standard uncertainty as the lowest common denominator for all uncertainty calculations, irrespective of their source, type or statistical distribution. If we reduce all elemental uncertainties to their standard uncertainty, we can combine them to find the standard uncertainty of a measurement. Standard uncertainty does not have a probability associated with it.

The expanded uncertainty is found by applying a probability to the standard uncertainty. The most common way is to multiply the standard uncertainty with a coverage factor.

Root Sum of the Squares (RSS)

If we wish to combine several numbers, we could just add them up. For example, adding 3, 6 and 10 we get 19. However, in uncertainty analysis if we use a simple addition it will typically lead to an overestimate of the combined uncertainty. To resolve this, we extensively use a root sum of the squares, or RSS, calculation. For the above example, the RSS is:

$$\sqrt{(3)^2 + (6)^2 + (10)^2} = 12.04$$

The Big Picture

When we take a measurement we use a transducer. This may be a simple device such as a ruler, or a complicated electronic device with inbuilt signal conditioning and analysis. It does not matter what transducer is used, the resulting measurement has some uncertainty. This uncertainty has two sources: Uncertainty in the transducer itself, and inherent variability in the quantity being measured, the measurand.

When a calculated quantity is determined from one or more measurements the uncertainty in the final quantity is determined by propagating the uncertainty.

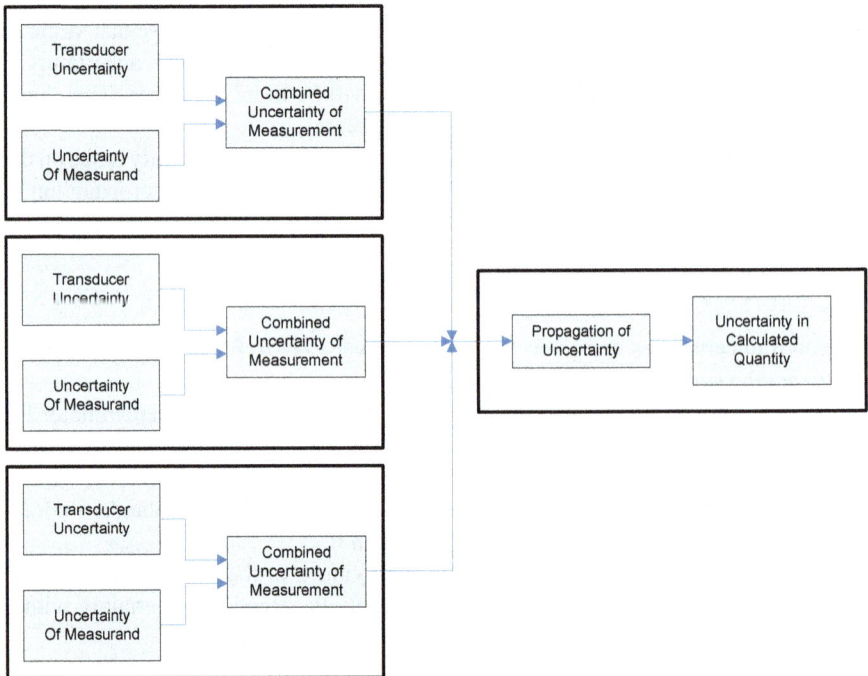

Type A and Type B Elemental Uncertainties

This chapter discusses elemental uncertainties and their origin. There is no mathematics in this chapter. The next chapter deals with the mathematics.

Elemental uncertainties are those associated with a particular measurement. As such, they are related to the accuracy of measured quantities, rather than the accuracy of calculated quantities. Elemental uncertainties are inherent in every transducer, with better quality transducers usually having less uncertainty (i.e., they are more accurate) than lower quality transducers. Elemental uncertainties may also be due to variability in the measurand. For example, if you are measuring the diameter of a shaft using calipers, the calipers have a certain 'accuracy.' Also the shaft diameter will vary slightly from place to place due to size variations. These are both examples of elemental uncertainties. For calibrated transducers the calibration certificate will include the effect of all elemental uncertainties associated with the transducer itself. While it can be educational to look into the cause for the transducer uncertainty, the transducer uncertainty will not need to be calculated since the calibration certificate gives everything that is need. However, there will still be additional uncertainty in the measurand that needs to be included in the uncertainty analysis.

There are fundamentally two types of elemental uncertainty: those that are repeatable and those that vary. The variable uncertainties were previously called random uncertainties; they vary from measurement to measurement; from time to time; and from place to place. Conversely, the common wisdom is that repeatable uncertainties do not vary. We will soon see that this is not actually the case!

Random Uncertainty

If it is possible to identify which statistical distribution (for example, normal distribution, uniform distribution, U-distribution) best models the random uncertainties it may be possible to identify the properties of the distribution, such as mean, standard deviation, skewness and kurtosis. Using these properties we can predict a range in which the next random event is likely to occur, but we are not able to predict the actual value of that event—this is the very nature of randomness.

Calibration can estimate (quantify) the amount of transducer random elemental uncertainty, but the calibration cannot reduce it.

One way of reducing the effect of randomness is to take a large number of readings and calculate the average. Statistically, this is reducing the standard deviation of the mean by increasing the sample size. As an example, random uncertainties are those that cause a bathroom scale to show a slightly different reading each time you step off and back on again; averaging several readings will help reduce this variability.

If you can only take a single measurement, typically the only way to reduce transducer elemental uncertainty due to randomness is to have a better quality transducer. Randomness in the measurand is usually reduced by, for example, higher quality manufacturing of the item being measured. Sometimes environmental issues can introduce randomness. For example, ground vibration may affect the accuracy with which a theodolite can be aimed at a target, and electrical interference will affect electronic apparatus. The amount of random uncertainty can be reduced by isolating the transducer (physically or electronically) from the environment. None of these approaches can completely eliminate random elemental uncertainty.

Repeatable, or Systematic, Uncertainty

As was mentioned previously, the common wisdom is that systematic uncertainties are repeatable and do not vary. However this is not true. Systematic uncertainties can vary significantly from measurement to measurement. The systematic uncertainty will be the same if all measurement conditions are identical, but for different measured values or under different measuring conditions, the uncertainty can be different.

Improving the quality of equipment can reduce systematic uncertainty. Calibration can also reduce the elemental uncertainty, but only (at best) to the uncertainty of the calibration process. Systematic uncertainty can never be totally eliminated.

Averaging repeated measurements will not change the systematic uncertainty. Indeed, since systematic uncertainties are repeatable their effect is not easy to see, and one can easily be lead to the false conclusion that taking the average of several measurements will give a number that is close to the true value. *Just because you can get an experiment to give repeatable results does not mean it is accurate*!

Systematic uncertainty can be reduced by improved calibration and better quality equipment, but it can never be eliminated totally. Thus, for a given type of transducer, elemental uncertainty can only be reduced by using a more sensitive and accurate transducer. This can be expensive. An uncertainty budget can help determine which transducers in a measurement scenario need to be improved. The topic of uncertainty budgets is discussed later in this monograph.

Systematic and random uncertainties are demonstrated in the following figure:

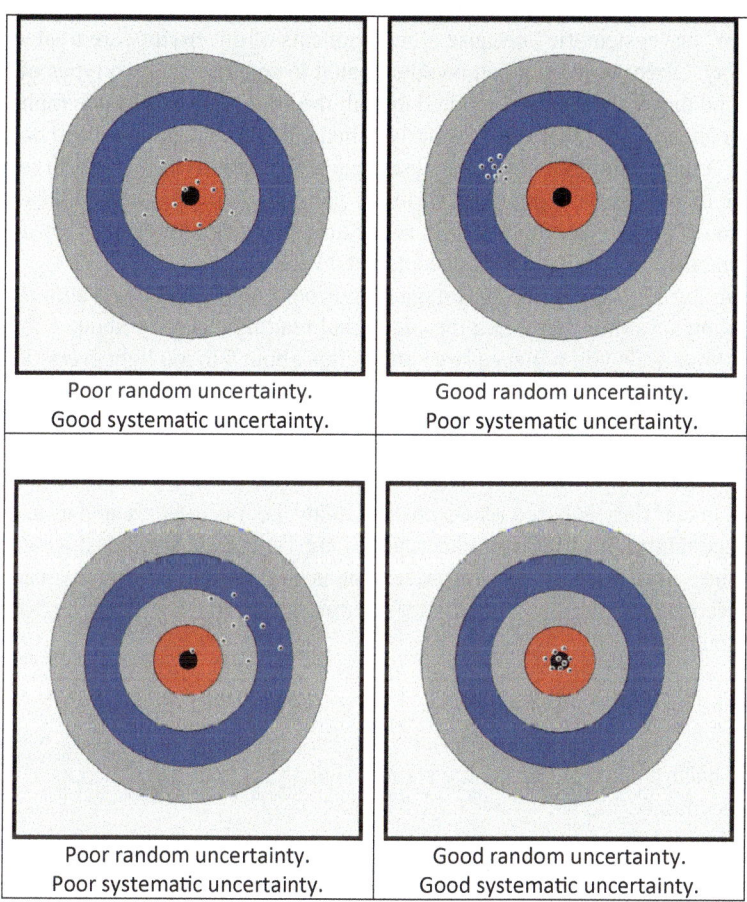

| Poor random uncertainty. Good systematic uncertainty. | Good random uncertainty. Poor systematic uncertainty. |
| Poor random uncertainty. Poor systematic uncertainty. | Good random uncertainty. Good systematic uncertainty. |

Type A and Type B Elemental Uncertainty

As mentioned previously, the GUM—the Guide to Uncertainty in Measurement—was first published in 1992 by ISO/BIPM. It has seen several revisions and is now accepted in industry as the recognized method for determining uncertainty in all measurements. The GUM classifies elemental uncertainties in two categories: Type A and Type B. The GUM requires that uncertainties that can be evaluated by statistical methods be treated as Type A uncertainties, and those evaluated by any other means be treated as Type B uncertainties.

As described in the GUM, the uncertainty analysis is based on the concept that there is no inherent difference between an uncertainty component arising from a random effect and one arising from a correction for a systematic effect. As a consequence, and as stated in the GUM, it is unnecessary to classify components as "random" or "systematic" because all components of uncertainty are treated in the same way. Often, though, it can be educational to separate the two types of uncertainty, and that is the method adopted in both the GUM and this monograph.

Most uncertainties that were formerly called random uncertainties will be treated as Type A uncertainties. This is because we use the underlying statistical model to estimate them. Most uncertainties formerly called systematic uncertainties will be treated as Type B uncertainties. This is because a systematic 'error' does not have any variability, and thus it cannot be analyzed with statistics.

Returning to your bathroom, imagine your bathroom scale has a zero offset so that it shows 5 lb too light, and the individual readings vary by about 1 lb. When you use your scale you will measure your weight about 5 lb too light *every time*. The zero offset is systematic and is therefore a Type B uncertainty. The variability from reading to reading of about 1 lb is a Type A uncertainty.

> If an uncertainty is based on a statistical analysis, it is to be treated as a Type A uncertainty. Most random uncertainties are Type A uncertainties.
>
> They vary each time a measurement is made. The uncertainty can be reduced by averaging lots of measurements, but it can never be totally eliminated.

> Any uncertainty not based on a statistical analysis is to be treated as a Type B uncertainty.
>
> Most systematic uncertainties are Type B uncertainties.

> Both Type A and Type B elemental uncertainties can be reduced by having a better quality transducer.
>
> Calibration can quantify Type A elemental uncertainty, but not reduce it.
>
> Calibration can reduce the level of Type B elemental uncertainty but there will always be a residual amount that cannot be determined.

The one exception to the above rule of classifying uncertainties as Type A or B based on statistics is the treatment of the uncertainty introduced by *scale resolution*. There are some good arguments as to why it should be treated as a Type A uncertainty, for example a statistical uniform random distribution is used to assess this uncertainty. There are also some good arguments as to why it should be treated as a Type B uncertainty. For this specific case the GUM states that *scale resolution is to be treated as a Type B uncertainty*.

Sources of Elemental Uncertainty

Let's look at some of the more common sources of elemental uncertainty.

Repeatability Repeatability is the ability of a transducer to give the same output when it is used several times to measure the same thing.

Repeatability is considered as randomly distributed with a normal distribution. It is a Type A uncertainty.

Thermal Stability Many systems, both mechanical and electrical, can be sensitive to temperature. For example, the output from a strain gage depends on the resistance of the metal foil wires that make the gage. The wires' resistance depends on temperature, and if a gage is used in a changing temperature environment the output caused by the temperature change can wrongly be attributed to a changing strain. While the effect of large temperature changes can be mollified by techniques such as temperature compensation, there will be some residual non-compensated effect.

Thermal stability is often treated as a random Type A uncertainty with a normal distribution.

Noise By "noise" we do not (normally) mean acoustic noise, although there are some measurements that are sensitive to acoustic noise. Rather, we usually mean the effect on a signal due to electrical interference from surrounding electrical and magnetic fields.

Noise is usually treated as a random Type A uncertainty.

Resolution, Scale Size and Quantization Most devices do not give continuous output; rather, the output is in the form of a series of steps. For example, when you use a simple tape measure to measure a length, you might quote the length to the nearest 1/8 in. Wire-wound potentiometers are limited to the change in resistance caused by the pick-up moving over a discrete coil. Many digital systems include analog-to-digital conversion, which automatically introduces the "stepped" output, and digital displays are limited to the resolution of the least significant digit.

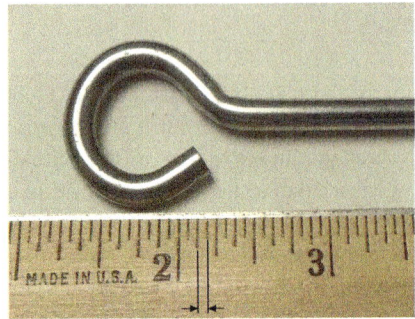

Resolution is treated as a Type B uncertainty.

Hysteresis Hysteresis causes a reading to be different depending on whether the device is being "loaded" or "unloaded" when the measurement is taken. As an example, the hysteresis in a spring balance is caused by mechanical problems such as friction and bearing misalignment in the device. Thus, hysteresis will cause a mechanical scale to consistently measure the applied weight too low (for increasing load) or consistently too high (for decreasing load). Other types of device also demonstrate hysteresis although the hysteresis in electrical devices is usually very small. There are some systems (such as spectrum analyzers) that have user-selectable variable hysteresis.

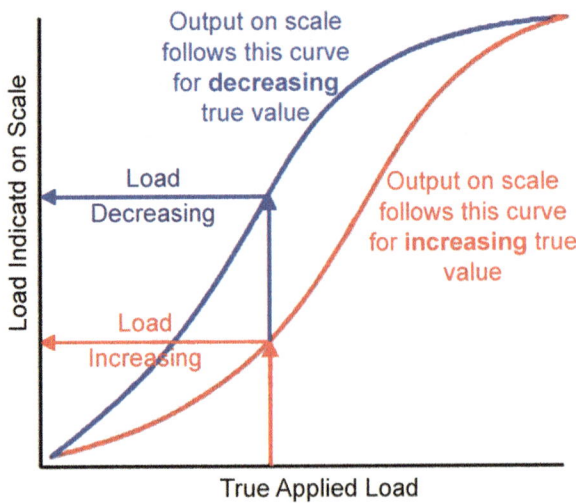

Sources of Elemental Uncertainty

Hysteresis cannot be analyzed by statistics and it is classified as a Type B uncertainty.

Common Mode Voltage When different voltages are applied to two input terminals of an amplifier, the amplifier will produce an output. Ideally, if the *same* voltage (relative to ground) is applied to the terminals the amplifier will produce zero output. Common mode voltage uncertainty is the uncertainty caused by the amplifier actually producing some output under these conditions.

Common mode voltage is treated as a Type B uncertainty.

Installation A pitot tube can be used to measure air speed, for example in aircraft. An example of installation uncertainty is when a pitot tube is removed and replaced. Because of the boundary layer effect, if the pitot tube is not put back in *exactly* the same place the reading will be slightly different for the same true value of air speed. Another example is using calipers to measure the diameter of a sphere. Slight misplacing of the calipers from the true diameter can result in a measurement that is slightly too small.

Installation is treated as a Type B uncertainty.

Nonlinearity (or Linearity) This is a common problem for calibration. It is often assumed that doubling the input to a transducer will double its output. For many transducers this assumption may be adequate, however, actual nonlinearity of the transducer will cause the measurement to have a systematic uncertainty. One extreme example of nonlinearity is the thermocouple, where the voltage output is not a linear function of the temperature. In an attempt to take into account the inherent nonlinearity of the thermocouple, systems designed to measure temperature with thermocouples incorporate polynomials in the conversion from voltage to temperature. Since the polynomials do not *exactly* resolve the issue there is still some residual systematic uncertainty.

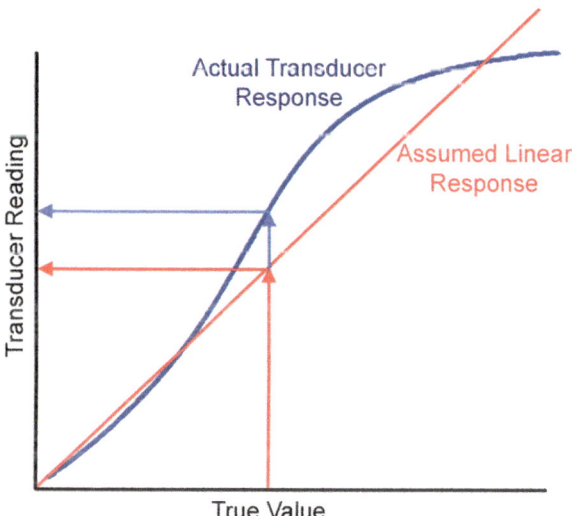

Nonlinearity is treated as a Type B uncertainty.

Spatial Variation We have already hinted at uncertainty due to spatial variation when, in the introduction to this monograph, we said that the temperature in a room would vary from place to place. If we were measuring near a "hot spot", the thermometer would *always* read too high. Spatial variation is typically associated with the measurand (rather than the transducer). As an example, if a shaft has a nominal diameter of 2 in its actual diameter will vary slightly from place to place, and this variability is systematic—repeatedly measuring the shaft at the same place will determine the same diameter.

Spatial variation is a Type B uncertainty.

Loading Imagine putting a cold mercury-in-glass thermometer into a beaker of hot water—that is, if you can still find a mercury thermometer and your safety department will let you use it! Otherwise, try the experiment with an alcohol-in-glass thermometer! Some of the heat will go from the water into the thermometer. As a result, the final temperature of the hot water will be too low because the water has cooled down a bit. This is an example of a loading issue. Many transducers cause loading uncertainty. As another example, imagine that during quality control you have to use a micrometer to measure the thickness of corrugated cardboard. The micrometer squeezes the cardboard and the thickness you measure depends upon how firmly you grasp the cardboard. The ratchet in the micrometer is to enable a more consistent jaw pressure, but there is still some variability from measurement to measurement.

Loading uncertainties are treated as Type B.

Zero Offset This is often caused if a device is not "zeroed" properly. That is, when the device does not give a zero reading when the quantity being measured is zero. Calibration and setting the zero point both aim to minimize this uncertainty, but they never totally remove it.

Zero offset is a Type B uncertainty.

Sensitivity This is the measure of how much the output of a transducer varies as the input (the measured quantity) varies. For example, the quoted sensitivity of an accelerometer may be 98.1 mV/g. This indicates that the device will generate a 98.1 mV signal if the input is 1 × (acceleration due to gravity). While calibration aims to give the best sensitivity for a transducer, there will still be some residual error.

Unknown errors in sensitivity cause systematic uncertainty, and they are treated as Type B uncertainties.

> Uncertainties that are random in nature can usually be analyzed with statistics, and are treated as Type A uncertainties.
>
> Systematic uncertainties are repeatable and cannot be analyzed with statistics. Consequently they are normally treated as Type B uncertainties.

Accuracy—Pandora's Box?

Accuracy is defined as how close the measurement is to the true value. Although we use the term *accuracy*, it is really the *inaccuracy* that is specified. Different manufacturers have different interpretations of their meaning of the term *accuracy*, and in a real-world application you should be careful to ensure you are using the appropriate definition. However, when accuracy is quoted, it normally includes all the residual Type A and Type B uncertainties in the measuring system. Accuracy is often quoted as the percentage of full scale. Thus for a balance that can weigh up to 250 lb with the accuracy quoted as 1% of full scale, the uncertainty is ±2.5 lb regardless of the reading or divisions on the scale.

Despite the fact that some companies quote a transducer's accuracy, NIST (Technical Note 1297, 1994 edition) states, "Because 'accuracy' is a qualitative concept, one should not use it quantitatively, that is, associate numbers with it."

Final reminder

Remember that there are many more sources of elemental uncertainty. If you need to classify them, apply the logic that if the analysis of the uncertainty can be done with statistics it is a Type A uncertainty. If statistics cannot be used (for example, if the uncertainty is systematic and repeatable) then it is to be treated as a Type B uncertainty.

If you wrongly classify a Type A elemental uncertainty as Type B, or vice versa, the consequences may not be significant. While the mistake will change the attribution of uncertainty to different components of the uncertainty analysis, the final estimated uncertainty will be the same, irrespective of wrong Type A/B classification!

> Wrongly classifying Type A and Type B uncertainties will lead to different intermediate numbers.
> But the final uncertainty will be the same!

Standard Uncertainty of a Measurement   3

Recall that elemental uncertainties are those that are either due to uncertainty in the measurement transducer or due to variability in the parameter being measured, and so far we have looked at some sources of these uncertainties. In this chapter we will develop a method for numerically combining the elemental uncertainties, resulting in the standard uncertainty of a particular measurement.

But standard uncertainty isn't everything. In the next chapter we will see how to take the standard uncertainty of a measurement we develop here, and expand it to determine the *expanded uncertainty of a measurement*, or, as it is often called, "the uncertainty."

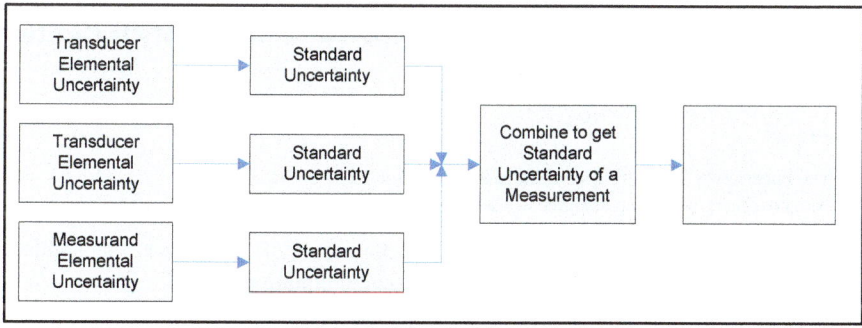

A critical concept with the Type A and B elemental uncertainty analysis is standard uncertainty, given the symbol lower case u. We can think of the standard uncertainty as the lowest common denominator for all uncertainties. We reduce all elemental uncertainties to their standard uncertainty. We then combine these standard uncertainties to find the standard uncertainty of measurement. Optionally (and usually), we can then develop the expanded measurement uncertainty, as we will see in the next chapter.

There are several advantages of using the standard uncertainty approach. Primarily, it usually leads to a better, more realistic, estimate of the uncertainty of a measurement than was possible prior to the standard uncertainty methods developed in the GUM. Another advantage of the standard uncertainty approach is that elemental uncertainties with different levels of confidence can be combined after they have been reduced to their standard uncertainty. This was not easily possible with the older bias and precision methods which are no longer in use.

Also, if an (expanded) uncertainty is known at one level of confidence, it can quickly be converted to a different level of confidence; you can quickly determine a 99% confidence uncertainty if you already know the 95% confidence uncertainty.

> If you are working on a project that specifies Bias and Precision you are using an outdated method that will most likely give an unrealistic estimate of the final uncertainty.
>
> Learn the Type A and Type B methods in this monograph, and impress your boss with your up-to-date knowledge!

All uncertainties, irrespective of whether they are Type A or Type B, are individually established as their standard uncertainty. The key to the analysis, therefore, is to know how to determine a standard uncertainty from the given information. The following sections give the procedure for determining the standard uncertainty for a variety of possible situations.

Finding the Standard Uncertainty when an Uncertainty Is Quoted at a Certain Level of Confidence

Examples:

The uncertainty in speed is ±5 mph at 95% confidence,
At a coverage of 2, the resistance is ±5 Ω.

If the elemental uncertainty is quoted at a certain level of confidence, the standard uncertainty u can be calculated from the following equation.

$$\text{Standard Uncertainty}, u = \frac{\text{(uncertainty)}}{k}$$

The k-factor, or coverage factor, is the number of standard deviations from the mean that include the percentage confidence. When an uncertainty is quoted at a level of confidence, the underlying statistical distribution is often the normal distribution, in which case the k-factors are given in the table below. While theoretical statistics can determine different k-factors for (for example) 94 and 95% confidence, the estimation of most uncertainties is, itself, an approximation. Therefore, it does not make sense to try to distinguish between 94 and 95%. This is especially the case if working at a confidence level above 99%, where the tails of the theoretical normal distribution are rarely good estimates of reality. Therefore, the table 3.1 includes

Table 3.1 Coverage factors (k-factors) for different levels of confidence (normal distribution)

Confidence (%)	Significance (%)	Theoretical k-factor	Approximate k-factor
68.27	31.72	1.000	
70	30	1.036	1
80	20	1.282	1.3
90	10	1.645	1.6
95	5	1.960	2
95.45	4.55	2.000	
98	2	2.326	
99	1	2.576	3[a]

[a] 99% confidence is well into the tails of the normal distribution, and therefore sensitive to the shape of the true distribution. NIST suggests using a k-factor of 3 when the confidence is required as "99% or more"

approximations that can often be adopted without significant loss of credibility in the uncertainty analysis.

If the underlying distribution is not a normal distribution, the statistics of the relevant distribution could be used to determine the appropriate coverage factor. However, when several elemental standard uncertainties have been combined, the k-factor for the normal distribution should still be used, even if the distributions for the individual uncertainties are not normal. This is, in part, due to the statistical central limit theorem: "*Given certain conditions, the arithmetic mean of a sufficiently large number of iterates of independent random variables, each with a well-defined expected value and well-defined variance, will be approximately normally distributed.*" Rice (1995), Mathematical Statistics and Data Analysis (Second ed.), Duxbury Press.

Note that confidence, significance and k-factor are three different ways of saying essentially the same thing. A confidence of 95% means we are "*95% confident that the true value of the measurand is within the quoted uncertainty.*" This is the same as saying there is a "*5% chance of a significant event*," the significant event being that the true value is outside the uncertainty range. Both of these statements are encompassed by stating the uncertainty is given for a k-factor of 2.

Example: The uncertainty in a temperature measurement T is given as $\pm 3.5°F$ at 20% significance.
The standard uncertainty is $u_T = \dfrac{3.5}{1.3} = 2.69°F$.

Finding the Standard Uncertainty when an Elemental Uncertainty Is Given as a Standard Deviation with No Information About the Sample Size

Examples:

The average measured pressure was 2580 psi with a standard deviation of 43 psi.
"The variance in length was 0.04 in^2." Note that variance = (standard deviation)^2. Thus for this example, standard deviation= $\sqrt{0.04} = 0.2$in.

When the elemental uncertainty is given as a standard deviation, and there is no information about the sample size, the standard uncertainty is taken to be the same as the given standard deviation

$$\text{Standard Uncertainty}, u = (\text{Given standard deviation})$$

Example: The standard deviation of the temperature measurement T was 3.5° F.
The standard uncertainty is $u_T = 3.5°\text{F}$.

Finding the Standard Uncertainty when the Uncertainty Is Given as a Standard Deviation and the Sample Size Is also Given

Examples:

Twenty measurements of pressure had an average of 2580 psi with a standard deviation of 43 psi.
The variance of 0.04 in^2 was determined from a sample of five length measurements.

Often in this situation the standard uncertainty is taken to be the same as the given standard deviation. However, this can lead to an underestimate of the standard uncertainty. The following presents a statistically consistent method for finding the standard uncertainty.

First the standard deviation is used with the sample size and the statistical Student-t distribution to determine the 95% confidence interval.

Then, the calculated 95% confidence interval is converted to a standard uncertainty using a k-factor of 2. These two steps are combined in the following single equation.

$$\text{Standard Uncertainty}, u = \frac{t_{95\%, dof} \times (\text{Given standard deviation})}{2}$$

Table 3.2 Values of Student-t for selected degrees of freedom and levels of confidence

Degrees of Freedom (dof)	1.3	1.6	2.0	3.0	k-factor
	20%	10%	5%	1%	Significance
	80%	90%	95%	99%	Confidence
1	3.07768	6.31375	12.70620	63.65674	
2	1.88562	2.91999	4.30265	9.92484	
3	1.63774	2.35336	3.18245	5.84091	
4	1.53321	2.13185	2.77645	4.60409	
5	1.47588	2.01505	2.57058	4.03214	
10	1.37218	1.81246	2.22814	3.16927	
15	1.34061	1.75305	2.13145	2.94671	
20	1.32534	1.72472	2.08596	2.84534	
25	1.31635	1.70814	2.05954	2.78744	
50	1.29871	1.67591	2.00856	2.67779	
100	1.29007	1.66023	1.98397	2.62589	
Infinity	1.28157	1.64488	1.96001	2.57593	

In this equation $t_{95\%, dof}$ is the Student-t value for 95% confidence (2-sided distribution) with *dof* mathematical degrees of freedom. For experiments that calculate the normal and standard deviation from *n* measured data,

$$\text{Degrees of Freedom, } dof = n - 1$$

While for a least squares analysis of *n* data pairs that determines parameters for both slope and intercept,

$$\text{Degrees of Freedom, } dof = n - 2$$

Student-t values are tabulated below for a selection of levels of confidence and degrees of freedom (Table 3.2).

The Student-t values can also be calculated in many computer packages. The following are functions for Excel, Matlab and Mathcad, programs that are often used for uncertainty analysis. If you wish to test your coding skills, set the uncertainty to 95% (usually entered as 0.95 in most computer programs) and the degrees of freedom to 10. The calculated Student-t should be 2.228139.

Excel These examples assume that cell A1 contains the required confidence (0.95) and cell A2 holds the degrees of freedom (10).

$= \text{TINV}(1-A1, A2)$	This function is maintained for backward compatibility. Microsoft warns it may be removed from future releases of Excel.
$= \text{T.INV.2T}(1-A1, A2)$	This function calculates the inverse of the two-sided Student-t distribution. It is a simple replacement of the TINV function and is available from Excel 2010.
$= \text{T.INV}(0.5+A1/2, A2)$	The T.INV function calculates the inverse of the one-sided Student-t distribution. Note that 0.95 probability has to be entered into the function as (0.5+0.95/2) in order to obtain the 2-sided statistic required for uncertainty analysis.

Matlab The Matlab function TINV calculates the inverse of the one-sided Student-t distribution. Comparable to the Excel T.INV function, care has to be taken when entering the probability. The following is a short Matlab script that will calculate the Student-t value required for the uncertainty analysis. If you are not familiar with Matlab, the % symbol is a comment symbol. Everything on a line after the % is ignored by the interpreter.

```
p=0.95; % required confidence, entered as percent/100
dof=10; % remember that dof is usually (# measurements-1)
t=TINV(0.5+p/2,dof); % the Student-t required for uncertainty analysis
```

Mathcad Mathcad is a mathematical package that uses typeset mathematical notation on a graphical user interface. The following partial screen shot shows the use of Mathcad's qt() function to calculate the Student-t value.

$$p: = 0.95$$
$$d: = 10$$
$$qt\left(0.5+\frac{p}{2}, d\right) = 2.228$$

Example: Based on sixteen measurements of the pressure P in a pressure vessel, the standard deviation is 15.2 psi.

Student-t is found for 95% confidence and 16−1=15 degrees of freedom. From the table, $t_{95\%,15} = 2.13145$

The standard uncertainty is

$$u_P = \frac{t_{95\%,15} \times (\text{StdDevn})}{2} = \frac{2.13145 \times 15.2}{2} = 16.199 \text{psi}$$

Finding the Standard Uncertainty when the Elemental Uncertainty ...   25

Further information about the Student-t distribution can be found in most standard statistics text books.

If you look at the table of Student-t values you will see that except for small sample sizes, Student-t (at 95% confidence) is close to 2, and the calculated standard uncertainty is close to the given standard deviation. In some industries it is common practice just to use the given deviation as the standard uncertainty, irrespective of sample size. This typically leads to an underestimate of the uncertainty.

> Example: Based on a large number of measurements of the pressure P in an experiment, the standard deviation is ±5 psi.
> Student-t for a "large number" of measurements is about 2. Therefore the standard uncertainty is
> $$u_P = 5 \text{ psi}$$

Finding the Standard Uncertainty when the Elemental Uncertainty has a Known Statistical Distribution

Not all elemental uncertainties follow the normal distribution. Some of the more common examples are shown in the following table, where the standard uncertainty is the same as the theoretical standard deviation of the distribution.

Elemental uncertainty	Discussion	Standard uncertainty, u
Most random elemental uncertainties	Normal distribution defined by standard deviation, σ. See above if the sample size is small.	$u = \sigma$
Quantization with ROUNDING (See below for special notes)	Uniform distribution, with resolution a. For example, if the smallest increment on a scale is 1/8 in, then $a = 0.125$ in and $u = 0.03608$ in	$u = \dfrac{a}{\sqrt{12}}$ $u = \dfrac{(a/2)}{\sqrt{3}}$
Quantization with TRUNCATION (See below for special notes)	Digital displays often truncate readings. Uniform distribution, with resolution a. For example, if the smallest digit on a scale represents 0.01 lb, then $a = 0.01$ lb and $u = 0.00577$ lb	$u = \dfrac{a}{\sqrt{3}}$
A-to-D conversion	The quantization error in voltage is uniformly distributed. If an A/D convertor has n bits, and is set to a full scale signal input range of V_{MAX} volts, the resolution is $V_{MAX}/2^n$. A/D converters can *round* or *truncate* values. Simple A/D usually truncate. More complex systems can round	$u_{ROUND} = \dfrac{V_{MAX}}{2^n \sqrt{12}}$ $u_{TRUNCATE} = \dfrac{V_{MAX}}{2^n \sqrt{3}}$

Elemental uncertainty	Discussion	Standard uncertainty, u
Triangular distribution	A triangular distribution is proposed by the GUM for cases where all values must fall within the range $\pm a$ and there is a central tendency for values	$u = \dfrac{a}{\sqrt{6}}$
R.F. mismatch measurements	RF power incident on a load may be delivered to the load with a phase angle between $-\pi$ and π radians and the probability of occurrence between these limits is uniform. If we assume that the amplitude of the signal is sinusoidal, the distribution for incident voltage is a U-shaped distribution limited to $\pm a$	$u = \dfrac{a}{\sqrt{2}}$

Example: An analog pressure gage has divisions that are in 25 psi increments. Values are recorded to the nearest division (rounding).
The standard uncertainty is

$$u = \frac{a}{\sqrt{12}} = \frac{25}{\sqrt{12}} = 7.21688 \text{ psi}.$$

Example: The analog speedometer in a car has divisions that are in 5 mph increments. The driver interpolates to estimate his speed to the nearest 1 mph.
The standard uncertainty due to the resolution is

$$u = \frac{a}{\sqrt{12}} = \frac{1}{\sqrt{12}} = 0.2887 \text{ mph}$$

The driver gets pulled over for speeding. He wants to "down play" his speed and truncates to the next lowest 5 mph division.
The standard uncertainty is now

$$u = \frac{a}{\sqrt{3}} = \frac{5}{\sqrt{3}} = 2.887 \text{ mph}$$

Example: A truncating 8-bit analog-to-digital convertor is set to a maximum input of 10 V.
The standard uncertainty is

$$u = \frac{V_{MAX}}{2^n \sqrt{3}} = \frac{10}{2^8 \sqrt{3}} = 0.02255 \text{ V}$$

Special Notes on Scale Resolution For those who are not faint of heart and welcome a mathematical challenge, detailed derivations of the uncertainty due to scale size, quantization and analog-to-digital conversion are given in the appendix. For those who just want to get the job done, the results are summarized here.

When calculating the standard uncertainty due to scale resolution it is important to recognize the difference between *full resolution and half resolution*. Different people understand resolution to be different things. Some people claim resolution is the smallest increment which a numerical display can indicate (the scale size). Others argue that the reading will never be more than half of the scale size from the indicted value (think here of using a ruler to measure a length), and therefore the resolution is *half* of the smallest increment. How you personally want to define resolution is your concern. Not wanting to bicker over something that ultimately makes no difference to the uncertainty analysis, in this monograph we compromise and refer to the first case as "full-resolution" and the second case as "half-resolution."

It is also important to recognize the difference between rounding and truncation. For most analog scales the readings are typically *rounded* to the nearest scale increment. Conversely most digital systems *truncate* the readings. For example if the scale size on a thermometer is 1 °F a value of 23.6 °F will be rounded to 24 °F but truncated to 23 °F. As a second example, if the scale size on a balance is 20 lb a value of 496 lb will round to 500 lb and truncate to 480 lb.

Full-resolution with rounding (applicable to most analog transducers). If the smallest divisions on a ruler are at 1 mm intervals, the full-resolution is 1 mm. The standard uncertainty is then calculated as:

$$u = \frac{\text{(full-resolution)}}{\sqrt{12}} = \frac{1\text{mm}}{\sqrt{12}} = 0.289\text{mm}$$

Half-resolution with rounding (applicable to most analog transducers). A measurement made using the same ruler with 1 mm intervals will be quoted to the nearest division, in which case the error caused by scale size will not be more than 0.5 mm (half of the 1 mm scale interval). This quantity is the half-resolution and the standard uncertainty is calculated as:

$$u = \frac{\text{(half-resolution)}}{\sqrt{3}} = \frac{0.5\text{mm}}{\sqrt{3}} = 0.289\text{mm}$$

Providing care is taken to match the type of quoted resolution (full or half) with the correct equation (divide by $\sqrt{12}$ or $\sqrt{3}$), the resulting standard uncertainty is the same.

Digital Display with Truncation. The standard uncertainty is related to the smallest interval that is shown on the display. For example, if the smallest "division" on a digital balance is 0.01 g. The standard uncertainty is:

$$u = \frac{\text{(display increment)}}{\sqrt{3}} = \frac{0.01\text{g}}{\sqrt{3}} = 0.00577\text{g}$$

Truncation with error adjustment. If your display truncates its readings, the average displayed value is half a scale increment lower than the true value. If you add this systematic offset (error) to every measurement, the uncertainty is then the same as when we have rounding

For example, the temperature displayed on a digital display is truncated to 0.1 °F. If you measure 78.4°F, the standard uncertainty due to truncation is:

$$\frac{a_{SCALE}}{\sqrt{3}} = \frac{0.1}{\sqrt{3}} = \pm 0.0577 \,°F$$

However, adding the systematic uncertainty (half of the scale size) means that you would use the number 78.45°F as the measured value and the standard uncertainty is given by the rounding equation:

$$u_{ADJUSTED} = \frac{a_{SCALE}}{\sqrt{12}} = \frac{0.1}{\sqrt{12}} = \pm 0.0289 \,°F$$

Analog-to-digital (A/D) conversion. Simple A/D converters can either round or truncate, depending upon the firmware in the converter. More advanced systems use much more elegant data capture and signal processing algorithms that do not directly relate to rounding or truncation, but can be treated as if they round.

If the A/D converter maximum input is set to V_{MAX} and the converter has n bits, the standard uncertainty is:

$$\text{with rounding, } u = \frac{V_{MAX}}{2^n \sqrt{12}}$$

$$\text{with truncation, } u = \frac{V_{MAX}}{2^n \sqrt{3}}$$

If in doubt about the internal working of the A/D converter, assume that it truncates, since this gives a larger standard uncertainty. Afterwards, if your analysis shows that the A/D is a critical component of the overall measurement uncertainty, you probably need to address this aspect in your process design.

When using an analog (rounded) scale:

$$\text{Standard uncertainty} = \frac{\text{(full-resolution)}}{\sqrt{12}} = \frac{\text{(half-resolution)}}{\sqrt{3}}$$

When using a digital display (with truncation):

$$\text{Standard uncertainty} = \frac{\text{(smallest display increment)}}{\sqrt{3}}$$

When using a digital display (with truncation and systematic error adjustment):

$$\text{Standard uncertainty} = \frac{\text{(smallest display increment)}}{\sqrt{12}}$$

When using an analog-to-digital converter with n bits set to a maximum input of V_{MAX}:

$$\text{with rounding, standard uncertainty} = \frac{V_{MAX}}{2^n \sqrt{12}}$$

$$\text{with truncation, standard uncertainty} = \frac{V_{MAX}}{2^n \sqrt{3}}$$

Combining Several Standard Uncertainties to find the Standard Uncertainty for a Single Measured Quantity

So far we have taken a number of different elemental uncertainties and converted each one to its standard elemental uncertainty. The next step in the process is to combine all these standard uncertainties.

The method recommended in the GUM is to first categorize all the uncertainties as Type A or Type B. The Type A and Type B combined standard uncertainties are then calculated separately as follows:

$$\text{Type A combined standard uncertainty} = u_A = \left\{ \sum_i u_{A,i}^2 \right\}^{1/2}$$

$$\text{Type B combined standard uncertainty} = u_B = \left\{ \sum_i u_{B,i}^2 \right\}^{1/2}$$

In other words, the Type A combined standard uncertainty is calculated as the root sum of the squares (RSS) of the individual Type A standard uncertainties, and similarly for the Type B uncertainties.

The *combined* standard uncertainty for the measurement is then determined as:

$$\text{Standard uncertainty of the measurement} = u_m = \left\{ u_A^2 + u_B^2 \right\}^{1/2}$$

Example

A mass standard is being used to calibrate a balance. Estimate the standard uncertainty of measurement based on the following elemental uncertainties. For this exercise let us assume that these are the only uncertainties pertaining to the problem, although in the real world there are many more.

The raw signal from the balance is digitized using a 16-bit analog-to-digital converter set to a maximum input of 10 g. This signal is then presented on a display with a resolution of 0.01 mg. The 95% uncertainty due to nonlinearity of the balance is estimated at 0.0070 mg and repeatability testing gave a standard deviation of 0.0060 mg based on 21 observations. Continuous monitoring of the balance identified that the signal noise had a 99% confidence interval of 0.004 mg.

First, classify each uncertainty as Type A or Type B.

Uncertainty	Discussion	Type A or B
Analog-to-Digital converter	GUM states quantization and resolution are to be treated as Type B	B
Resolution of the display	GUM states quantization and resolution are to be treated as Type B	B
Nonlinearity	Systematic	B
Repeatability	Random in nature. Also, the information quoted is a standard deviation. Therefore it is based on a statistical analysis.	A
Signal noise	Random in nature.	A

For each elemental uncertainty, determine its standard uncertainty.

Uncertainty	Discussion	Elemental standard uncertainty (mg)
Analog-to-Digital converter (truncation assumed) (Type B)	Careful with units (1 g = 1000 mg): $$u_{A/D} = \frac{a}{2^n \sqrt{3}} = \frac{10 \times 1000}{2^{16} \sqrt{3}}$$	0.088097
Resolution of the display (truncation assumed) (Type B)	$$u_{RES} = \frac{a}{\sqrt{3}} = \frac{0.01}{\sqrt{3}}$$	0.005774
Nonlinearity (Type B)	Quoted at 95% confidence, so use a k-factor of 2. $$u_{NONLIN} = \frac{U}{k} = \frac{0.0070}{2}$$	0.0035
Repeatability (Type A)	Given as standard deviation with 21−1=20 degrees of freedom. From the Student-t table, $t_{95\%,20} = 2.08956$. $$u_{REPEATABILITY} = \frac{t_{95\%,15} \times (\text{StdDevn})}{2}$$ $$= \frac{2.08596 \times 0.0060}{2}$$	0.006258

Combining Several Standard Uncertainties to find the Standard Uncertainty ...

Uncertainty	Discussion	Elemental standard uncertainty (mg)
Signal noise (Type A)	Quoted at 99% confidence, so use a k-factor of 3. $$u_{NOISE} = \frac{U}{k} = \frac{0.004}{3}$$	0.001333

Separately for the Type A and Type B uncertainties, find their combined standard uncertainty using an RSS calculation.

$$\text{Type A:} \: u_A = \left\{ \sum_i u_{A,i}^2 \right\}^{1/2} = \left\{ (0.006258)^2 + (0.001333)^2 \right\}^{1/2} = 0.006398 \, \text{mg}$$

$$\text{Type D:} \: u_B - \left\{ \sum_i u_{B,i}^2 \right\}^{1/2} - \left\{ (0.088097)^2 + (0.005774)^2 + (0.0035)^2 \right\}^{1/2}$$
$$= 0.088355 \, \text{mg}$$

Now combine these two uncertainties to find the combined standard uncertainty of the measurement. Again, use an RSS calculation:

$$\text{Standard uncertainty of the measurement} = u_m = \left\{ u_A^2 + u_B^2 \right\}^{1/2} = 0.88586 \, \text{mg}$$

Expanded Uncertainty of a Measurement and an Uncertainty Budget for a Single Measurement

4

In the previous chapter we saw how to determine the standard uncertainties for a variety of elemental uncertainties. We also saw how to combine these standard uncertainties to determine the standard uncertainty for a measurement.

Standard uncertainties are the lowest common denominator, and can be added (using an RSS—Root Sum of the Squares calculation) and compared with relative ease. However, they do not directly tell us anything about the confidence we should have in a particular reading. We will see in this chapter how we can take the standard uncertainty of a measurement and expand it to determine the *expanded uncertainty of a measurement*, or, as it is often called, just "the uncertainty."

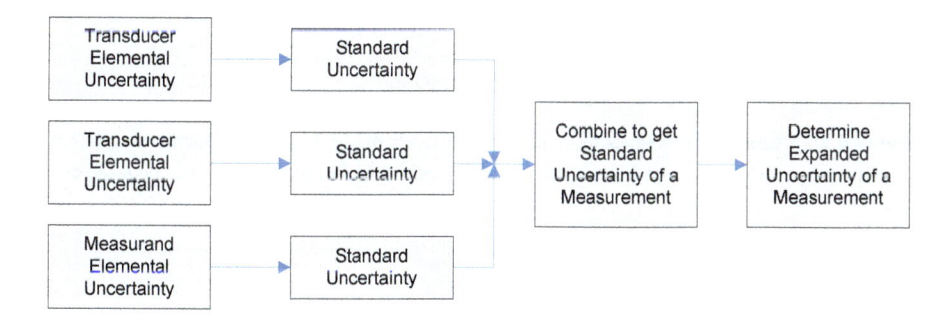

The expanded uncertainty is always quoted with a statistical confidence. As an example (the full explanation and mathematics follow later in this chapter), the standard uncertainty of a temperature reading might be 2.5 °F. We will find that if we can accept an uncertainty of ±7.5 °F in our measurement, then we will be 99% certain of measuring within requirements.

In this chapter we also start to look at uncertainty budgets. These are reports of uncertainty that can help identify, for example, where a process needs attention or where cost savings can be made with no effect on the final product. Later (in the next chapter and in the case study) we develop different uncertainty budgets, from relatively simple to extensive.

Expanded Uncertainty, U

The expanded uncertainty for a measurement is given the symbol upper case U. (Be careful to make upper and lower case U's clearly different in your work since lower case refers to standard uncertainty, and upper case refers to expanded uncertainty. These are the symbols recommended by the GUM.)

The expanded uncertainty of a measurement, U_m, depends upon a chosen level of confidence, and is calculated as:

$$\text{Uncertainty of the measurement} = U_m = k.u_m$$

> Lower case for standard uncertainty, u
> Upper case for expanded uncertainty, U
> with
> $$U = k.u$$

We use the same k-factor as shown previously and repeated in the table below. Unless specifically required otherwise, it is common practice to use 95% confidence, or a k-factor of 2 (Table 4.1).

Table 4.1 Coverage factors (k-factors) for different levels of confidence (normal distribution)

Confidence (%)	Significance (%)	Theoretical k-factor	Approximate k-factor
68.27	31.72	1.000	
70	30	1.036	1
80	20	1.282	1.3
90	10	1.645	1.6
95	5	1.960	2
95.45	4.55	2.000	
98	2	2.326	
99	1	2.576	3[a]

[a] 99% confidence is well into the tails of the normal distribution, and therefore sensitive to the shape of the true distribution. NIST suggests using a k-factor of 3 when the confidence is required as "99% or more"

Example

In the last chapter we developed the standard uncertainty of measurement in an example concerning the calibration of a balance. The example is repeated here:

> The raw signal from the balance is digitized using a 16-bit analog-to-digital converter set to a maximum input of 10 g. This signal is then presented on a display with a resolution of 0.01 mg. The 95% uncertainty due to nonlinearity of the balance is estimated at 0.0070 mg and repeatability testing gave a standard deviation of 0.0060 mg based on 21 observations. Continuous monitoring of the balance identified that the signal noise had a 99% confidence interval of 0.004 mg.

In the last chapter we found that the standard uncertainty of measurement for this balance was $u_m = 0.088586$ mg.

We now wish to estimate the 95% (expanded) uncertainty of measurement. From the table of k-factors we see that the appropriate factor for 95% coverage is $k=2$.

$$\text{Uncertainty of the measurement} = U_m = k.u_m = 2 \times 0.088586 = 0.177173 \text{ mg}$$

Final Statement: The 95% uncertainty of measurement for the balance is ± 0.18 mg.

Changing the Level of Confidence

Imagine you owned this balance and rather than you having to calculate the measurement uncertainty, its calibration sheet gave you the uncertainty as ± 0.18 mg at 95% confidence. However, you need the uncertainty with 99% confidence. Other than sending the balance in for recalibration, what could you do?

The solution is rather easy. First, using the given 95% uncertainty we calculate the standard uncertainty using a k-factor of 2.

$$U_m = k.u_m$$

$$0.18 = 2 \times u_m$$

$$u_m = \frac{0.18}{2} = 0.090 \text{ mg}$$

We now use a k-factor of 3 (from the table for 99%) to find the new uncertainty of measurement:

$$U_m = k.u_m = 3 \times 0.090 = 0.27 \text{ mg}$$

Final Statement: The 99% uncertainty of measurement for the balance is ± 0.27 mg.

An Uncertainty Budget

Imagine that you have been tasked to improve the accuracy of the balance. You will do this by trying to reduce one of the uncertainties. Which one do you address first, and why? Which elemental uncertainty does not need further inspection? These are the kinds of questions that an uncertainty budget can help resolve. In essence, an uncertainty budget is a table of uncertainty values. Budgets can have minimal information (much like the one we develop here), or they can have significant detail. The next chapter includes an example of an uncertainty budget that includes a cost analysis. Separately, the full worked case study develops three different levels of uncertainty budget for a problem, and the differences between them are discussed.

The essence of an uncertainty budget is that it presents a tabular comparison of the uncertainties associated with a measurement or process. Depending upon the specific project, an uncertainty budget table can include many different columns, each being part of the uncertainty analysis. When the budget is for a single measurement (as in this example), the table should, as a minimum, include a column of the numerical elemental standard uncertainties. For uncertainty budgets for a more complete propagation the table should include at least the individual expanded uncertainties of measurement.

Uncertainty budgets can be presented using standard uncertainties, or they can be presented using expanded uncertainties at a chosen level of confidence. The choice is yours! However, all uncertainties should be shown in similar fashion. It is not permissible to mix standard uncertainties and 95% confidence expanded uncertainties in the same table.

The following table gives one example of an uncertainty budget for the balance calibration problem above. All of the numbers are extracted from the numerical solution presented in this and the previous chapter. Since we use RSS to combine the elemental uncertainties, we choose to include a column to compare the square of each standard uncertainty. This is not essential or required for every uncertainty budget—we just chose to do it in this case.

Source	Elemental standard uncertainty (mg)	(Uncertainty)^2 (mg)^2
Analog-to-Digital converter	0.088097	7.761×10^{-3}
Resolution of the display	0.005774	33.33×10^{-6}
Nonlinearity	0.00350	12.25×10^{-6}
Repeatability	0.006258	39.2×10^{-6}
Signal noise	0.001333	1.78×10^{-6}

We see that the uncertainty from the analog-to-digital conversion is extremely large when compared with all the other uncertainties. Therefore we should focus our improvement efforts here. Recall that we assumed this A/D truncated values, and this may have led to a slight over-estimate of its uncertainty. However, since it transpires that the A/D contribution is the major contributor to the measurement uncertainty by a factor of more than 200, we need to address this component in the design of the balance.

Maybe you could use an A/D convertor with more than 16 bits, perhaps one with 24-bit which is currently the gold standard in instrumentation. This single change would significantly reduce the 95% uncertainty of measurement from ±0.178 down to ±0.019 mg.

Alternatives might be to reduce the maximum A/D input (if this is possible) which was set at 10 g, or redesign the balance to eliminate this A/D conversion altogether if possible.

The second highest uncertainty is repeatability. For mechanical balances this is often related to friction in the system, so better quality components such as linkages, bearings and springs might be needed, although the overall improvement would be extremely small. If all of the uncertainty due to repeatability could be eliminated, the 95% confidence uncertainty of measurement when quoted to two significant figures, 0.18 mg, would not change.

It is pointless trying to improve electronic shielding to reduce the signal noise (smallest standard uncertainty). Even if you could remove the signal noise in its entirety, this would only reduce the final 95% uncertainty of measurement from 0.177173 to 0.177153 mg, a trivial change that represents no real improvement.

So based upon our analysis of the uncertainty budget, if we wish to improve the uncertainty of this balance the only place it is worth investing improvement funds is in the analog to digital conversion aspect.

Summary of the Steps to Determine the Uncertainty of Measurement

1. Determine the standard uncertainty for each elemental uncertainty, u.
2. Separately for the Type A and Type B uncertainties combine the standard uncertainties with an RSS calculation to get the total Type A u_A and Type B u_B standard uncertainties.
3. Use an RSS calculation to combine the total Type A and Type B uncertainties to get the total standard uncertainty in a measurement.
4. Multiply the standard uncertainty in measurement by a k-factor to get the expanded (total) uncertainty of measurement U, for a prescribed level of confidence.

Propagation of Uncertainty & An Uncertainty Budget Example

The earlier chapters of this monograph dealt with the uncertainties in individual transducers and measurements. If you are using just one transducer to measure something (for example, a balance to weigh a component), the uncertainty methods and analysis we have already developed will give you the desired uncertainty in your measurement.

However, in engineering it is common to take one or more measurements and use them to *calculate* the quantity of interest. For example, you may measure the diameter of a shaft and from this measurement calculate the radius. Or you may make several measurements of different types of quantities (for example, the diameter of a ball and its mass) and use them to calculate the density of the object. In both of these examples it is necessary to determine the uncertainty of the calculated quantity based on the uncertainties of each individual measurement.

This process is called propagating the uncertainty. In this chapter we will study how to propagate the uncertainty from several measured quantities in order to determine the uncertainty in a calculated quantity. Remember that first you have to estimate the uncertainty of each separate measurement using methods from the previous chapters. Then use the methods in this chapter to combine uncertainties for several measurements.

It is important to recognize the difference between a measured quantity and a calculated quantity. As we mentioned earlier, you may talk loosely about "measuring the area of a piece of paper." In actuality, you most probably mean that you *measured* the length and width, and then *calculated* the area. This distinction between measured and calculated is critical to a correct uncertainty analysis.

Examples of measured and calculated quantities:

> Diameter of a shaft, $D = 2.46$ in   MEASURED QUANTITY
> Radius of the shaft, $r = \dfrac{D}{2} = 1.23$ in   CALCULATED QUANTITY

> Note that even something as fundamental as obtaining the radius of a shaft by measuring its diameter and dividing by 2 is a calculation, and forces the need to propagate uncertainties. If you measure the diameter of a shaft with calipers with an uncertainty of ± 0.002 in, the uncertainty in the calculated radius is *not* ± 0.002 in.

> Temperature, $T_F = 68\,°\text{F}$ MEASURED QUANTITY
> Temperature, $T_C = \dfrac{(T_F - 32)}{1.8} = 20\,°\text{C}$ CALCULATED QUANTITY

> Something as apparently simple as converting between different units requires a calculation. If the uncertainty in T_F is $\pm 2\,°\text{F}$, the uncertainty in T_C is also $\pm 2\,°\text{F}$, but it just does not make sense to mix units and say a temperature is $20\,°\text{C} \pm 2\,°\text{F}$. We would rather say the temperature is $20\,°\text{C} \pm 1.1\,°\text{C}$. Units matter!

> Beam width, $b = 15$ mm   MEASURED QUANTITY
> Beam thickness, $t = 3$ mm   MEASURED QUANTITY
> Second Moment of Area $= I = \dfrac{bt^3}{12}$
> $= 33.75 \text{ mm}^4$ CALCULATED QUANTITY

> Hot body temperature, $T_1 = 600$ K   MEASURED QUANTITY
> Cold body temperature, $T_2 = 300$ K   MEASURED QUANTITY
> Radiation Heat Transfer, $q = 5.669 \times 10^{-8} \times 0.8 \times \left(T_1^4 - T_2^4\right)$
> $= 5510 \text{ W/m}^2$ CALCULATED QUANTITY

Loads added to a piano wire and the resulting extensions	MEASURED QUANTITIES
Diameter of the wire, gage length	MEASURED QUANTITIES
Regression analysis to find Young's modulus of elasticity	CALCULATED QUANTITY

As demonstrations of the propagation of uncertainty, we will consider two examples: We will use the tip deflection of a rectangular cross section cantilever to determine the material's elastic modulus, and we will calculate the radiation heat flow from a hot body to a cooler body based upon measurements of temperature. While these particular measurement processes may not be relevant to you, together they demonstrate the principals necessary for many measurement processes just by changing the equations and numbers.

We also develop an uncertainty budget example to help assess financial aspects of a measurement process.

General Principles

In general, the quantity we are calculating is a function of several measured quantities. As an example, consider a cantilever (for example, a beam that is secured to a bench at one end, and hangs over the edge of the bench at the other end). If we hang a load on the free end of the cantilever the tip will deflect downwards. We can use the relationship between applied load and the deflection of the tip to determine the elastic modulus, or Young's modulus of elasticity, for the material the cantilever is manufactured from. This method is certainly not an acceptable or approved process for finding the elastic modulus of a material, but it is useful for demonstrating the propagation of uncertainty methods.

Using standard mechanics of materials equations for the elastic deflection of thin beams, we can determine the elastic modulus, E, as a function of several measured quantities:

$$E = \frac{4mgL^3}{ybd^3}$$

where m is the mass hung on the tip, L, b and d are the length, breadth (width) and depth of the beam respectively, and y is the observed tip deflection.

When we say we want to propagate the uncertainty, what we mean is that if there is a small uncertainty in, for example, the measurement of the thickness of the beam, we want to find out how much this affects the calculated quantity (the elastic modulus in this example). The uncertainty in each measured quantity will affect the result a different amount. We will assume that each measured quantity can be in error by a small amount δ. We can find the propagated uncertainty in the elastic modulus δ_E from the following partial differential equation.

Note that in this example we have assumed that we know the acceleration due to gravity, g, exactly, although this isn't really the case. If you have a situation that might be sensitive to the actual acceleration due to gravity, we direct you to the accompanying case study where we investigate the uncertainty in the acceleration due to gravity in detail.

$$\delta_E = \delta_m \frac{\partial E}{\partial m} + \delta_L \frac{\partial E}{\partial L} + \delta_y \frac{\partial E}{\partial y} + \delta_b \frac{\partial E}{\partial b} + \delta_d \frac{\partial E}{\partial d}$$

The terms in the partial differential equation have real significance. For example, the term $\frac{\partial E}{\partial m}$ is the *sensitivity* of E with respect to the mass. In other words, if we change the mass a little bit, how much does this change the calculated value of E? We will see later in this chapter that we can use experimental or numerical methods to estimate many of these sensitivities if the mathematics is too complicated for partial differentiation.

If C is a calculated quantity, and q is a measured quantity, the partial differential equation $\frac{\partial C}{\partial q}$ is called the sensitivity of C with respect to q.

The sensitivity tells us how much the calculated quantity changes if there is a small change in the measured quantity.

The units associated with each sensitivity depend upon the units of C and q. For example, if the calculated quantity is a pressure quoted in psi, and the measured quantity is a length measured in inches, the sensitivity will be in psi/inch.

Note that we could show the units as $lb/inch^3$, but for sensitivities we do not normally do this. We usually show them as

(calculated units)/(measured units).

The above partial differential equation is only mathematically exact if all of the δ's are infinitesimally small; otherwise it is an approximation. We will make the assumption that the uncertainties of the measured quantities are all small and replace all the δ's in the equation with the expanded uncertainties, U, for each measured

Absolute and Relative Uncertainty

value. We also note that the sensitivities can be positive or negative, and the uncertainties themselves carry a ± attribute. Without care, mathematically this can incorrectly lead to the possibility that one uncertainty might apparently "cancel out" another uncertainty. In reality they don't. Therefore we need to ensure that the plus/minus signs associated with both the sensitivities and the uncertainties are chosen correctly.

There are several strategies for compensating for the different signs. The method given in the GUM determines a quantity often called the *best estimate of uncertainty*, which combines the propagated uncertainties using an RSS (Root Sum of the Squares) approach:

$$U_E = \left\{ \left(U_m \frac{\partial E}{\partial m} \right)^2 + \left(U_L \frac{\partial E}{\partial L} \right)^2 + \left(U_y \frac{\partial E}{\partial y} \right)^2 + \left(U_b \frac{\partial E}{\partial b} \right)^2 + \left(U_d \frac{\partial E}{\partial d} \right)^2 \right\}^{1/2}$$

Note that all the individual uncertainties of measurement, U_m, U_L, etc., must be quoted to the same level of confidence. If the individual uncertainties of measurement are given with different levels of confidence, use the method given in a previous chapter to change them to be at the same level of confidence. Unless a project dictates otherwise, it is common practice to quote all uncertainties at 95% confidence (5% level of significance, or a k-factor of 2).

Absolute and Relative Uncertainty

Absolute uncertainty is quoted in the same units as the measured quantity. The U_m, U_L, U_y, etc. in the above equation are all absolute uncertainties. As an example, if an electrical current is quoted in milliAmperes, the absolute uncertainty will also be quoted in milliAmperes.

$$\text{Current } I = 500 \text{mA}$$

$$\text{Absolute uncertainty in current } U_I = \pm 20 \text{ mA}$$

Relative uncertainty is a ratio and is either quoted in percentage or fractional form. It does not have units.

$$\text{Current } I = 500 \text{mA}$$

$$\text{Relative uncertainty in current} = \frac{\pm 20 \text{ mA}}{500 \text{ mA}} = \pm 0.04 = \pm 4\%$$

The absolute uncertainty, relative uncertainty and measured quantity are related by:

$$(\text{Relative uncertainty}) = \frac{(\text{Absolute uncertainty})}{(\text{Measured quantity})}$$

The words "absolute" and "relative" are not always used. Instead, it is often understood that uncertainties quoted in engineering units are absolute, and those quoted as fractions or percentages are relative. CORRECT USE OF UNITS IS IMPORTANT!

> We use ABSOLUTE uncertainty in the uncertainty analysis.
> If you have a relative uncertainty, you must first find the absolute uncertainty:
>
> $(\text{Absolute uncertainty}) = (\text{Relative uncertainty}) \times (\text{Measured quantity})$

Example: Determining Elastic Modulus with a Cantilever Beam We now use the cantilever example to demonstrate the propagation of uncertainty from measured quantities to calculated quantities.

A cantilever is $L=0.80$ m long, with an uncertainty of $U_L=\pm 0.005$ m. It is $b=35$ mm wide and $d=5$ mm thick, measured with uncertainties of $U_b=\pm 0.05$ mm and $U_d=\pm 0.005$ mm respectively. When a mass of $m=0.5\pm 0.01$ kg was hung on the tip, it was observed that the tip deflection was $y=11$ mm$\pm 3\%$. What are the elastic modulus E and the uncertainty in the modulus? All uncertainties of measurement are given to the 95% level of confidence. Assume the acceleration due to gravity, g, is known exactly.

Solution First calculate the elastic modulus. Note the following answer is given far too accurately. This excessive accuracy is required at this stage of the calculations, since we do not yet know how accurate the result is! After we have completed the uncertainty propagation we will know how accurately we can quote the final answer.

$$E = \frac{4mgL^3}{ybd^3} = \frac{4\times 0.5\times 9.81\times 0.8^3}{0.011\times 0.035\times 0.005^3} = 208.73642 \text{ GPa}$$

Note that all measurement uncertainties are quoted at 95% confidence and so we do not have to adjust any of them to a different level of confidence. However the tip deflection is given as a *relative uncertainty* of 3% so we need to calculate the *absolute uncertainty* U_y:

$$U_y = 0.011\times \frac{3}{100} = 0.33\times 10^{-3} \text{ m}$$

Next, we calculate all the sensitivity equations by finding the partial differential equations. If you have a calculus background you will find this mathematics isn't

Absolute and Relative Uncertainty

too bad—try it. If you don't have a calculus background, you may want to consult a friendly mathematician to help you!

For the main equation:

$$E = \frac{4mgL^3}{ybd^3}$$

The partial differential equations (sensitivities) are given in the following table, which also includes the measurement uncertainties from the given problem. The final column in the table calculates the contributions to the total uncertainty due to each propagated measurement uncertainty.

Quantity (x)	Measurement uncertainty, U_x	Sensitivity $\left(\frac{\partial E}{\partial x}\right)$	Contribution $\left(U_x \cdot \frac{\partial E}{\partial x}\right)$ (GPa)
Mass, m	0.01 kg	$\frac{4gL^3}{ybd^3} = 417.47$ GPa/kg	4.17
Length, L	0.005 m	$\frac{3 \times 4mgL^2}{ybd^3} = 782.76$ GPa/m	3.91
Tip deflection, y	0.33×10^{-3} m	$\frac{-4mgL^3}{y^2bd^3} = -18976$ GPa/m	−6.26
Breadth, b	0.05×10^{-3} m	$\frac{-4mgL^3}{yb^2d^3} = -5963.9$ GPa/m	−0.03
Depth, d	0.005×10^{-3} m	$\frac{-3 \times 4mgL^3}{ybd^4} = -125242$ GPa/m	−0.63

We can now find the uncertainty in E using an RSS calculation as:

$$\begin{aligned} U_E &= \left\{ \left(U_m \frac{\partial E}{\partial m}\right)^2 + \left(U_L \frac{\partial E}{\partial L}\right)^2 + \left(U_y \frac{\partial E}{\partial y}\right)^2 + \left(U_b \frac{\partial E}{\partial b}\right)^2 + \left(U_d \frac{\partial E}{\partial d}\right)^2 \right\}^{1/2} \\ &= \left\{ (4.17)^2 + (3.91)^2 + (-6.26)^2 + (-0.30)^2 + (-0.63)^2 \right\}^{1/2} \\ &= 8.511234 \text{ GPa} \end{aligned}$$

Our final solution is thus $E = 208.73642 \pm 8.511234$ GPa.

The uncertainty analysis tells us the accuracy of the measured quantity. Typically we would quote the final uncertainty to one or two significant figures, certainly not more. The calculated quantity is then rounded such that the least significant figure is quoted to the same order of magnitude as the uncertainty. For this example we chose to round the uncertainty to one significant figure: 8.511234 GPA rounds to 9 GPa. The calculated modulus of elasticity is then rounded such that its least significant figure has the same order of magnitude as the uncertainty giving our result. Thus $E = 209 \pm 9 \text{GPa}$.

The final statement must include a range and confidence. Thus the final statement could be any of the following. Note there are many more ways of giving the range and confidence!

The elastic modulus is 209±9 GPa with 95% confidence.
The elastic modulus is 209±9 GPa with 5% significance.
The elastic modulus is 209±9 GPa with a k-factor of 2.
At 95% confidence the elastic modulus is in the range 200–218 GPa.

> The propagated uncertainty should be round to one significant figure. Sometimes it can be rounded to two significant figures. Never more (as quoth the raven!).
>
> The calculated quantity is rounded such that its least significant figure is to the same order of magnitude as the (one significant figure) uncertainty.

> The final statement of the calculated quantity must include a range of values and the confidence in that range.

Uncertainty Budget—How to Use the Uncertainty Analysis to Improve the Accuracy of a Measurement Process

We can use the uncertainty analysis to set up an uncertainty budget. Essentially, this tabulates all the measured quantities along with the uncertainty they propagate to the final calculated quantity. The fully worked example included in the case study shows some different styles of uncertainty budgets, which can be used for a variety of situations including experimental and managerial. For example, it can become part of an overview of a manufacturing plant to help determine whether equipment needs calibration, repair or replacement. When final products are too expensive the uncertainty budget can help identify where savings can be made. Similarly, when it is necessary to improve the quality of a product the uncertainty budget can help identify the most cost-effective place to make changes.

For the cantilever example above, let us assume that the final uncertainty in the elastic modulus of ±9 GPa was deemed to be too large. What measurements should we change if we want to make the uncertainty smaller? Let us look at a summary of the propagated uncertainties which are extracted from the previous calculations and shown in the next table.

Quantity	Propagated uncertainty (contribution) at 95% confidence (GPa)
Added mass, m	4.17
Length, L	3.91
Tip deflection, y	6.26
Beam width, b	0.30
Beam thickness, d	0.63

We should try to improve the measurement that has the highest propagated uncertainty. In this case, it is the tip deflection with a contribution of 6.26 GPa. If making changes to reduce this uncertainty is too expensive or not technically possible, our next target measurement would be the added mass or the length measurement since both contribute about 4 GPa of uncertainty. However, neither of these changes would be as effective as improving the tip deflection measurement. Note that since we combine the contributions using a RSS calculation we should compare the (propagated uncertainties)² in the calculations; thus we are comparing the values $(6.26)^2$, $(4.17)^2$ and $(3.91)^2$, which are 39.2, 17.4 and 15.3.

The measurement that generated the lowest uncertainty in our experiment was the beam width, b, which introduced a propagated uncertainty contribution of 0.30 GPa. This is the least problematic measurement in our test, so putting a large effort into increasing the accuracy of this measurement is probably not cost effective. As a note, if we could completely eliminate the uncertainty associated with the beam width measurement, this would only reduce the final uncertainty in elastic modulus from ±8.511 down to ±8.506 GPa, which is essentially no improvement.

Uncertainty Budget Example: Radiation Heat Transfer

In this example we see how an uncertainty budget can help decide which equipment needs to be procured for an experiment. More expensive equipment will usually lead to smaller (better) uncertainty, but at what fiscal cost? The uncertainty budget can help you decide!

The problem we investigate is radiation heat transfer. When a hot body is close to a cold body, heat energy will radiate from the hot body to the cold body. The radiation heat transfer q (W/m²) between two bodies is given by:

$$q = c\left(T_1^4 - T_2^4\right)$$

The radiation heat transfer between the two bodies is a function of their absolute temperatures, T_1 and T_2, and a parameter, c. For a full analysis, the parameter c is dependent upon several other quantities, but for this uncertainty propagation example we will assume that it has been determined separately and independently.

For a particular pair of bodies let us assume that we know $c = 45.0 \times 10^{-9} \pm 0.2 \times 10^{-9}$ W/m²/K⁴, and prior knowledge tells us that the hot and cold temperatures will be about

$T_1 = 700$ K and $T_2 = 300$ K. In setting up a test you have a choice of two different thermocouples you could use to measure the temperatures. Complete with all purchase costs, installation costs and signal conditioning, "regular grade" thermocouples cost $ 125 each, and "best grade" thermocouples cost $ 200 each. According to the manufacturer, regular thermocouples can record a maximum temperature of 1300 °C with an uncertainty ±0.15 %, and best grade thermocouples can record a maximum temperature of 1700 °C with an uncertainty of ±0.5 °C.

You need one thermocouple for each temperature measurement. Determine the uncertainty versus cost for each possible thermocouple configuration.

Solution The actual heat transfer is calculated from the equation as 10,440.00 W/m².

Let's set up the required partial differential equations so that we can calculate the *sensitivities* needed for the uncertainty analysis. **The sensitivities do not change with type of thermocouple.** They only depend upon the equation used to determine the calculated heat transfer.

$$q = c\left(T_1^4 - T_2^4\right)$$

Parameter	Sensitivity wrt the parameter	Evaluated numerically
Coefficient c	$\dfrac{\partial q}{\partial c} = \left(T_1^4 - T_2^4\right)$	$232 \times 10^9 \; \dfrac{\text{W/m}^2}{\text{W/m}^2/\text{K}^4}$
High temperature T_1	$\dfrac{\partial q}{\partial T_1} = 4cT_1^3$	$61.74 \; \dfrac{\text{W/m}^2}{\text{K}}$
Low temperature T_2	$\dfrac{\partial q}{\partial T_2} = -4cT_2^3$	$-4.86 \; \dfrac{\text{W/m}^2}{\text{K}}$

Since the sensitivities have different units, we cannot compare them. For example we cannot say that 232×10^9 is bigger than 61.74. If we tried to compare values with different units it would be like saying an elephant is bigger than the color blue. (The authors believe that elephants are actually smaller than blue, but they have yet to prove it!) Yet again, units matter!

> Comparing sensitivities is rarely useful.
> They often have different units, so cannot be compared anyway.
> They need to be multiplied by their associated measurement uncertainty before they can be compared.

We now identify the *uncertainty of measurement* for each measured quantity/transducer. These uncertainties depend upon the transducers and measurands—**these uncertainties are independent of the sensitivities calculated above.** They just depend upon the instrumentation.

Uncertainty Budget Example: Radiation Heat Transfer

$$U_c = 0.22 \times 10^{-9} \text{ W/m}^2/\text{K}^4$$

$$U_{T-REGULAR} = \frac{0.15 \times 1300}{100} = 1.95°\text{C}$$

$$U_{T-BEST} = 0.5°\text{C}$$

> *Sensitivity equations* only depend upon the *equation* used to determine the calculated quantity. They do not depend upon the measurement uncertainties.
>
> *Measurement uncertainties* only depend upon the measurements—they do not depend upon the equation used to determine the calculated quantity or the sensitivities.

For this heat transfer example there are four different possible experimental configurations of thermocouples, and therefore four different uncertainty analyses we need to make:

- 2 × best grade thermocouples
- 1 × best grade thermocouple (high temperature) and 1 × regular grade thermocouple (low temperature)
- 1 × regular grade thermocouple (high temperature) and 1 × best grade thermocouple (low temperature)
- 2 × regular grade thermocouples

Let's consider each configuration in turn, and calculate the uncertainty in the calculated heat transfer. Remember that the sensitivities do not change for the different configurations since they are derived from the equation we used to calculate the heat transfer. Only the uncertainties for the individual measurements change as we "swap out" the different thermocouples.

The general equation we use for this problem is:

Uncertainty in heat transfer, U_q

$$= \left\{ \begin{array}{l} \left[(\text{Uncertainty in } c) \times (\text{sensitivity wrt } c)\right]^2 \\ + \left[(\text{Uncertainty in hot temp}) \times (\text{sensitivity wrt } T_1)\right]^2 \\ + \left[(\text{Uncertainty in cold temp}) \times (\text{sensitivity wrt } T_2)\right]^2 \end{array} \right\}^{1/2}$$

Using 2 × best grade thermocouples:

$$U_q = \left\{\left(U_c \frac{\partial q}{\partial c}\right)^2 + \left(U_{T_1} \frac{\partial q}{\partial T_1}\right)^2 + \left(U_{T_2} \frac{\partial q}{\partial T_2}\right)^2\right\}^{1/2}$$

$$= \left\{(0.2 \times 10^{-9} \times 232 \times 10^9)^2 + (0.5 \times 61.74)^2 + (-0.5 \times 4.86)^2\right\}^{1/2} = 55.78 \text{ W/m}^2$$

Using 1 × best (high temperature) and 1 × regular (low temperature):

$$U_q = \left\{\left(U_c \frac{\partial q}{\partial c}\right)^2 + \left(U_{T_1} \frac{\partial q}{\partial T_1}\right)^2 + \left(U_{T_2} \frac{\partial q}{\partial T_2}\right)^2\right\}^{1/2}$$

$$= \left\{(0.2 \times 10^{-9} \times 232 \times 10^9)^2 + (0.5 \times 61.74)^2 + (-1.95 \times 4.86)^2\right\}^{1/2} = 56.53 \text{ W/m}^2$$

Using 1 × best (low) and 1 × regular (high):

$$U_q = \left\{\left(U_c \frac{\partial q}{\partial c}\right)^2 + \left(U_{T_1} \frac{\partial q}{\partial T_1}\right)^2 + \left(U_{T_2} \frac{\partial q}{\partial T_2}\right)^2\right\}^{1/2}$$

$$= \left\{(0.2 \times 10^{-9} \times 232 \times 10^9)^2 + (1.95 \times 61.74)^2 + (-0.5 \times 4.86)^2\right\}^{1/2} = 129.0 \text{ W/m}^2$$

Using 2 × regular grade thermocouples:

$$U_q = \left\{\left(U_c \frac{\partial q}{\partial c}\right)^2 + \left(U_{T_1} \frac{\partial q}{\partial T_1}\right)^2 + \left(U_{T_2} \frac{\partial q}{\partial T_2}\right)^2\right\}^{1/2}$$

$$= \left\{(0.2 \times 10^{-9} \times 232 \times 10^9)^2 + (1.95 \times 61.74)^2 + (-1.95 \times 4.86)^2\right\}^{1/2} = 129.4 \text{ W/m}^2$$

Summary of the Results:

Radiation heat transfer, $q = 10{,}440$ W/m²				
High temp. Thermocouple	Low temp. Thermocouple	Cost ($)	Absolute uncertainty in q, U_q (W/m²)	Relative uncertainty in q, U_q (%)
Best	Best	400	55.8	0.53
Best	Regular	325	56.5	0.54
Regular	Best	325	129.0	1.24
Regular	Regular	250	129.4	1.24

Conclusions

Changing the low temperature thermocouple from "best" to "regular" makes very little difference to the final uncertainty. However, having a "best" thermocouple for the high temperature sensor makes a big difference (the uncertainty more than doubles if you use a regular grade thermocouple). Based solely on the uncertainty budget analysis, the following conclusions can be drawn:

- If 1.24% uncertainty is acceptable, it is cheapest to use two regular grade thermocouples.
- If 1.24% uncertainty is unacceptable, the cheapest solution is to have a best grade thermocouple measuring the high temperature, and a regular grade thermocouple for the low temperature.
- It is probably not worth the extra cost of installing a best grade thermocouple to measure the lower temperature.

Remember that these results are based solely upon the uncertainty of measurement, and while a difference of $ 75 between configurations may not seem significant, imagine the difference if you want to install an array of 25×25 thermocouples on each plate. This would make the difference in cost between experimental configurations nearly $ 50,000, which is significant to the authors! The results can help management make decisions, but the final decision has to take into account other factors. For example, the regular thermocouples might be on back order, and your project cannot wait. Or maybe if your boss's brother runs the company that makes the best grade thermocouples, guess which thermocouples you are likely to use, irrespective of your uncertainty budget analysis!

Earlier Examples Revisited

At the start of this chapter on the propagation of uncertainty we included several examples of measured and calculated quantities. For completeness we return to two of these examples and determine the propagated uncertainties.

Repeating the example of calculating the radius of a shaft from its diameter, earlier we gave:

$$\text{Diameter of a shaft, } D = 2.46 \pm 0.001 \text{in} \quad \text{MEASURED}$$
$$\text{Radius of the shaft, } r = \frac{D}{2} = 1.23 \text{in} \quad \text{CALCULATED}$$

5 Propagation of Uncertainty & An Uncertainty Budget Example

From this information,

$$\text{Uncertainty of measurement for the diameter, } U_D = 0.001 \text{in}$$
$$\text{where 95\% confidence is assumed}$$

$$\text{Sensitivity of radius wrt diameter} = \frac{\partial r}{\partial D} = \frac{\partial \left(\frac{D}{2}\right)}{\partial D} = \frac{1}{2} \text{in/in}$$

Since the diameter is the only measured quantity, the propagated uncertainty for the radius is calculated as:

$$U_r = \left\{ \left(U_D \frac{\partial r}{\partial D} \right)^2 \right\}^{1/2} = \left\{ \left(0.001 \times \frac{1}{2} \right)^2 \right\}^{1/2} = 0.0005 \text{in}$$

Thus, the uncertainty in radius is half of the uncertainty in diameter. In other words, the radius is *'two times more accurate'* than the diameter.

We now turn to the earlier example of converting Fahrenheit temperature readings to °C. For this exercise, let us assume that the 95% uncertainty of measurement is ±0.5°F.

$$\text{Temperature}, T_F = 68°F \pm 0.5°F \quad \text{MEASURED}$$
$$\text{Temperature}, T_C = \frac{(T_F - 32)}{1.8} = 20°C \quad \text{CALCULATED}$$

$$\text{Uncertainty of measurement in Fahrenheit, } U_F = 0.5°F$$

$$\text{Sensitivity of } T_C \text{ wrt } T_F = \frac{\partial T_C}{\partial T_F} = \frac{\partial \left(\frac{(T_F - 32)}{1.8} \right)}{\partial T_F} = \frac{1}{1.8} \text{°C/°F}$$

The propagated uncertainty for the temperature in °C is calculated as:

$$U_C = \left\{ \left(U_F \frac{\partial T_C}{\partial T_F} \right)^2 \right\}^{1/2} = \left\{ \left(0.5 \times \frac{1}{1.8} \right)^2 \right\}^{1/2} = 0.2778 °C$$

Thus, an uncertainty of ±0.5°F propagates to an uncertainty of ±0.28°C.

NEVER include any calculated quantities in the function. Even apparently simple sones like $A = bd$ or $r = D/2$.

Always propagate the uncertainties if there is *any* calculation involved.

Sensitivity by Perturbation

6

You have a problem. Your boss knows that you bought (and read!) this monograph and he now thinks of you as the company expert in uncertainty analysis. That's not the problem. The problem is that he believes the temperature of cooling fluid in your computer numerical control (CNC) milling machine is unduly affecting the size of the components. He wants to swap out the simple thermostat for a very expensive, computer controlled system a salesman is trying to sell him. The machine operator says it is nonsense, and you have been called in to arbitrate and help either improve product quality or save the company unnecessary expense. What do you do?

This is an example of where you probably do not need a full uncertainty analysis of the entire CNC process; rather you can just focus on the effect of cooling fluid temperature. Sadly there is no equation that says *"The size of the finished component is this function of cooling oil temperature."* Consequently, even though you are good at mathematics and can handle partial differential equations, with no governing equation you cannot find the sensitivity of component size with respect to the cooling fluid temperature. It is often the case in many engineering processes that the equations used to determine a calculated quantity are either not known or are intractable. In this chapter we investigate a perturbation method that often can help estimate some of the uncertainty sensitivities in these situations.

The method we develop can be used when you are faced with an actual measurement process (like this CNC mill example), or when the result is from a numerical simulation of the process. In either case we can apply perturbation methods to find the measurement uncertainty sensitivities. The concept of perturbation is relatively easy to grasp.

> By perturbation we mean that we perturb (make a small change to) one of the variables in the process, and see the effect it has on the output.

Let us divert for a moment to a different example we use to develop the concept of perturbation. Afterwards we will return to your CNC problem to see perturbation in action.

Imagine your company manufactures springs, and part of the process includes annealing the springs by immersing them in oil. Currently hot springs are put in a wire cradle and dropped into oil until the operator thinks they "are done." You are conducting an uncertainty analysis to see if the anneal time is significant to the final spring stiffness: Is the current approach adequate, or does the anneal time need to be controlled more closely? The physics behind the annealing cannot be modeled by a simple mathematical equation, so how do you progress?

First, you need to monitor and (accurately) measure the current process. Let us assume that you have identified the average anneal time $t_A = 90\,s$, with a measurement uncertainty (at 95% confidence) $U_{t_A} = \pm 8s$ Your next step is to set up a controlled experiment. In this case you will control the length of time the springs are annealed, and vary it slightly from trial to trial. In reality you would make several repeat trials, but for this example let us assume that you only conduct two tests. For the first test you controlled the anneal time to 89 s, and for the second test you ensured the anneal time was 94 s. After the springs have completed their manufacture process, you measure the spring stiffness (the output of the process) and find that the stiffness for the 89 s springs was 23.06 lb/in, and for the 94 s springs the stiffness was 23.04 lb/in.

What you did was to set up an experiment where you deliberately changed the anneal time by a small amount, and observed the change it caused in spring stiffness. This is a perturbation analysis.

So how do we use this information to get the measurement uncertainty sensitivities? For this example, we mathematically say that the spring stiffness S is some (unknown) function of anneal time t_A.

$$S = f(t_A)$$

We increase the anneal time by a small amount δt_A, and observe (measure) the resulting change in stiffness δS. From this information we can get an approximate numerical value for the rate of change of stiffness with respect to anneal time. Using the numbers from above (89 s gave 23.06 lb/in and 94 s gave 23.04 lb/in) we see that a 5 s increase in anneal time has caused a 0.02 lb/in reduction in stiffness. Mathematically:

$$\frac{\partial S}{\partial t_A} \cong \frac{\delta S}{\delta t_A} = \frac{-0.02}{5} = -0.004\,(\text{lb/in})/s$$

We use a perturbation experiment to determine the sensitivity of a calculated quantity with respect to one of the process inputs.

Thus, the sensitivity of stiffness to anneal time is approximately -0.004 (lb/in)/s.

> In the uncertainty analysis we replace the sensitivity determined from a partial differential equation with the sensitivity from the perturbation analysis.

We already know that the uncertainty in anneal time is $U_{t_A} = \pm 8 s$. We can now estimate the uncertainty this causes in spring stiffness:

$$U_S = U_{t_A} \frac{\partial S}{\partial t_A} \cong U_{t_A} \frac{\delta S}{\delta t_A} = 8 \times (-0.004) = -0.032 \text{ lb/in}$$

Thus, at 95% confidence the uncertainty in the annealing time results in an uncertainty of ± 0.032 lb/in in the spring stiffness. Is that significant or not? The uncertainty analysis cannot make that decision for you. It can only present the facts. If a stiffness uncertainty of ± 0.032 lb/in is acceptable, then current anneal timing methods are adequate. If this uncertainty is too high then the anneal time needs controlling more accurately.

For a more complete uncertainty analysis we should vary (perturb) every single measureable quantity separately, and observe the effect on the calculated property (output of the process). We would then combine these uncertainties using the methods in the earlier chapters to estimate the overall uncertainty in the final calculated quantity. However, it is very often the case in real-world applications that we only wish to see how much a single quantity affects the result.

Returning to the CNC Cooling Fluid Problem

So let's return to your CNC cooling fluid temperature problem. You investigate and find the following:

The existing temperature controller has a 95% uncertainty of $U_{T,EXISTING} = \pm 3\,°F$.

The proposed temperature controller has a 95% uncertainty of $U_{T,PROPOSED} = \pm 0.6\,°F$.

The component manufactured in the CNC has a total 95% uncertainty in size of a critical dimension D of $U_{D,EXISTING} = \pm 280.00 \times 10^{-6}$ in.

You run a perturbation experiment where you closely monitor the cooling fluid temperature, and observe the change in the size of the component. You conduct many trials and find:

If the cooling fluid's temperature is raised (perturbed) by $10\,°F$, you can measure a reduction in size of the component's critical dimension of 30×10^{-6} in.
This means that the sensitivity of product dimension to cooling fluid temperature is:

$$\frac{\partial D}{\partial T} \cong \frac{\delta D}{\delta T} = \frac{-30 \times 10^{-6}}{10} = -3.0 \times 10^{-6} \text{ in}/°F$$

The perturbation analysis has determined that the sensitivity of the critical dimension to cooling fluid temperature is -3.0×10^{-6} in/°F.

Now the uncertainty analysis! We know that the existing uncertainty in product size is due to uncertainty in temperature combined with the uncertainty of many other influences. We can summarize this as the following propagation of uncertainty RSS equation:

$$(\text{uncertainty in critical dimension}) = \sqrt{(\text{due to temperature uncertainty})^2 + (\text{due to other uncertainties})^2}$$

$$U_{D,EXISTING} = \sqrt{\left(U_{T,EXISTING}\frac{\partial D}{\partial T}\right)^2 + (\text{other uncertainties})^2}$$

Recall that the propagated contribution is given by:
(uncertainty in a measurement) × (sensitivity)
$U_{T,EXISTING}$ is the uncertainty in temperature (°F)
$\frac{\partial D}{\partial T}$ is the sensitivity of dimension w.r.t temperature (in/°F)
So the propagated uncertainty is

$$U_{T,EXISTING}\frac{\partial D}{\partial T}$$

Putting in the numbers for the existing situation:

$$280.0\times10^{-6} = \sqrt{\left(3\times3.0\times10^{-6}\right)^2 + (\text{other uncertainties})^2}$$

From which we calculate that $(\text{other uncertainties}) = 279.85\times10^{-6}$ in.

We now investigate what happens if we use the new temperature controller. The uncertainty can be found from:

$$U_{D,PROPOSED} = \sqrt{\left(U_{T,PROPOSED}\frac{\partial D}{\partial T}\right)^2 + (\text{other uncertainties})^2}$$

Putting in the numbers for the proposed temperature controller process:

$$U_{D,PROPOSED} = \sqrt{(0.6 \times 3.0 \times 10^{-6})^2 + (279.85 \times 10^{-6})^2} = 279.86 \times 10^{-6}$$

Thus we see that if we swapped to the new oil temperature computer controller we would reduce the 95% confidence uncertainty of the product's dimensions from $= \pm 280.00 \times 10^{-6}$ in to $= \pm 279.86 \times 10^{-6}$ in, and indeed improving the quality of a system component (the temperature controller) has improved the quality of the final component.

However, we normally, only quote the uncertainty to one (or at most, two) significant figures. Doing this we see that swapping the controller actually makes no realistic difference: The 95% uncertainty in product dimension stays at $\pm 280 \times 10^{-6}$ in, showing that replacing the temperature controller would not make a significant difference.

Don't forget to ask your boss for a bonus for saving the company a lot of money!

When we do not have a governing equation, or the mathematics is too complicated, we can identify the sensitivity of the calculated quantity using a perturbation of a measured quantity.

The perturbation can be a physical experiment, or it can be a numerical simulation—e.g., a finite element model or a simulation in Matlab

We then use this sensitivity in the uncertainty analysis.

Case Study

The case study offers an extended example which combines most of the aspects of an uncertainty analysis. It can be used as a template for the analysis of many different processes, not just the one presented in this study. The case study includes many of the problems that can hinder an analysis and gives suggested workarounds: Many measurements; insufficient measurements; mixed units; lack of calibration; human reaction time; acceleration due to gravity; unknown dimensions, and so on. The case study develops three different levels of uncertainty budget for the same process; from a relatively simple "per measurement" budget through to a detailed budget that not only identifies which transducers do or do not need calibrating, but also assigns the total process uncertainty to each aspect of elemental uncertainty associated with the transducers and measurands.

A Fully Worked Example Developing the Uncertainty Analysis for a Process, Including a Discussion of Uncertainty Budgets

This case study has been chosen because it demonstrates the application of most of the uncertainty analysis aspects that might be encountered with a measurement or process. The study is designed to act as a road map through the world of uncertainty. The specifics of this particular study will only be relevant to a few readers; however the approach and ideas are relevant to anyone needing to undertake an uncertainty analysis of a process.

The study considers using a Stormer viscometer to determine the viscosity of a fluid. As presented, the experiment is deliberately lacking in that the final uncertainty is too large for the results to be acceptable to the technical world. In this way, we can demonstrate how the uncertainty budgets can identify areas of weakness, and where improvements have to be made.

The Big Picture

This is an extensive example that incorporates much of the theory presented in the main body of this work. The general flow of the example is:

- Description of the experiment and the calculation required to determine the viscosity from the measured quantities.
- Presentation of sample measured data.
- Uncertainty of the transducers.
- Uncertainty of the measurements.
- Propagation of the measurement uncertainties to identify the uncertainty in viscosity (the calculated quantity).
- Uncertainty budgets—several different examples at different levels of detail are developed and discussed.

Standard Uncertainty or Expanded Uncertainty? One question that is often asked is whether the uncertainty calculations should be performed using standard uncertainties or expanded uncertainties. Recall that expanded uncertainties are calculated from standard uncertainties using a k-factor:

$$(\text{Expanded uncertainty}) = k \cdot (\text{standard uncertainty})$$
$$U = k \cdot u$$

The k-factor is 2 for 95% confidence, and other values for different levels of confidence are tabulated in the book.

It does not matter whether the calculations are conducted using standard or expanded uncertainties—you should get the same final uncertainty. However, it is not permissible to mix the two in any one calculation. To ensure consistency and avoid mistakes it is recommended that the following method is adopted:

- The initial part of the analysis is looking into individual transducer and measurand uncertainty. At this stage of the analysis use standard uncertainties and combine them to determine the standard uncertainty for each measurement.
- These standard uncertainties are then converted to expanded uncertainties for each measurement. That is, for each measurement multiply the standard uncertainty you have just calculated by a k-factor to determine the expanded uncertainty for the measurement.
- Propagate the expanded measurement uncertainties.

Spreadsheets Spreadsheets such as Excel and Google Docs are ideal for much of the analysis presented in this study. By making a judicious choice of layout, the final uncertainty budgets can appear as elegant, easy to read tables, with all the hard work being performed elsewhere in the spreadsheet. Corrections and additions can quickly be made to the analysis, with the final uncertainties automatically being calculated if changes are made.

While a spreadsheet is an excellent approach for a commercial application, the approach tends to "hide" the underlying analysis in the cells. Therefore, for the example presented here all calculations are shown in detail, and tables are restricted to only showing the information being discussed at the time.

An Excel Formatting Hint Have you ever had difficulty making Excel format numbers the way you want them to look? For example, if your number is 0.02156, Excel shows it as just that. What if you want to show it in exponential format? It might seem that using scientific format will be best. (Set the cell format by using a mouse right-click in the cell, select "format cells" from the dropdown menu, then pick scientific.) Sadly, this might not give you what you want:

| 2.16E-02 | You have lost the trailing digit, and Excel does not use 'engineering' format of keeping the exponent to multiples of 3 |

You can fix that! Rather than selecting scientific format select 'Custom'. In the box labeled 'Type' enter the following code. This makes Excel format numbers with exponents that are multiples of 3 and with 4 significant figures after the decimal point. Try this format: ##0.0000E+0. Try different variations to get your numbers looking just the way you want.

| 21.5600E-03 | Exponents are now multiples of 3, and there are 4 significant figures after the decimal point |

A Note on Significant Figures, Rounding, etc. The numbers presented in this worked example were all calculated in a spreadsheet. Thus full accuracy was carried forward. As a consequence of the rounding necessary for this typed version it may appear that some of the numbers are slightly inconsistent.

Despite the fact that uncertainty analysis can only identify *estimates* of the uncertainties, your interim calculations should always carry far more significant figures than might seem necessary, with rounding to appropriate significance only happening at the very end. Thus, conducting the calculations 'by hand' really is not recommended!

Description of the Experiment to Measure Viscosity

Viscosity arises due to internal friction between the molecules of a fluid whenever part of the fluid moves. Viscosity is thus the property of a fluid (liquid or gas) that is a measure of its resistance to shear. The Stormer viscometer, shown in the photograph and schematically in the figure, is an apparatus that puts a shearing motion into a fluid. The apparatus consists of a fixed outer cylinder and a rotating inner drum. The fluid that is being tested is contained in a thin layer between the cylinder and drum. The inner drum is driven by a falling weight and as the inner drum rotates this causes a shear stress in the fluid between the drum and cylinder. Once the drum reaches its terminal velocity the mass falls at a constant speed and the torque applied to the inner cylinder by the falling weight is balanced by the viscous resisting torque.

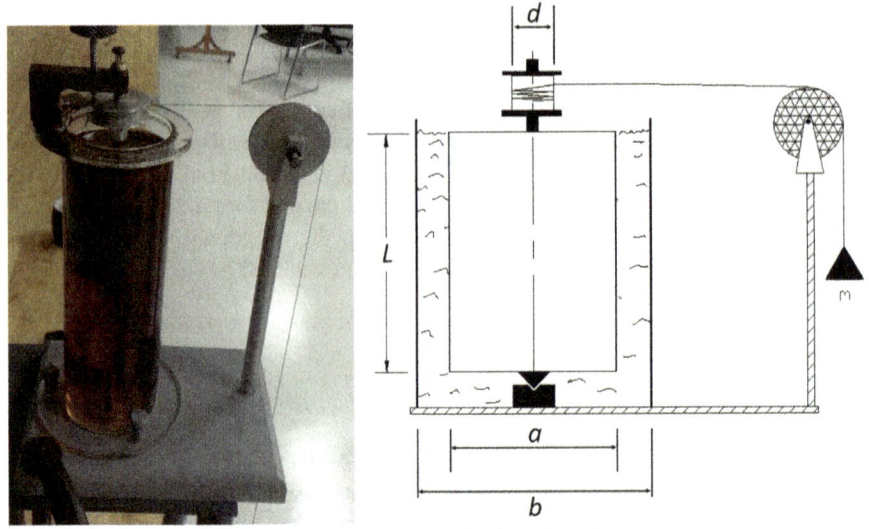

Stormer Concentric Cylinder Viscosity Apparatus

The theory pertaining to this experiment can be developed from the fundamental theories of fluid dynamics. However, for this uncertainty study a detailed knowledge of the theories and derivation of the equation is not necessary. This is often the case for uncertainty analysis.

> An uncertainty analysis can often be completed without a detailed knowledge of the underlying theory.
> The minimum that is needed is an equation that determines a calculated quantity from a series of measured quantities
> Better understanding of the theory may help in interpretation and redesign of an experimental configuration, but it should not change the results of an uncertainty analysis.

For this example the *calculated* quantity (the fluid's viscosity, μ) is a function of several *measured* quantities. The equation is:

$$\mu = \left[\frac{mgd^2(b-a)}{2\pi La^2 b}\right]\frac{t}{h} \tag{1}$$

Where:

Case Study

Symbol	Quantity
μ	Viscosity, N.s/m^2
a	Outside diameter of the rotating drum
b	Inside diameter of the fixed cylinder
d	Diameter of the pulley fastened to the top of the rotating drum
m	Mass suspended on the string
L	Length of the rotating drum (from fluid surface to bottom of drum)
t	Time it takes the mass to fall distance h
h	Distance mass falls at terminal velocity
g	Acceleration due to gravity

Measurements

We now detail the measurements obtained from a particular experiment.

Symbol	Units	Transducer	Measurements
a	mm	Calipers	Measured at several different locations: 52.75, 52.73, 52.85, 52.78, 52.72, 52.70, 52.74, 52.74, 52.77, 52.80, 52.78, 52.71, 52.81, 52.76, 52.81, 52.72
b	mm	Calipers	61.78
d	mm	Calipers	25.4
m	gram	Balance	100.3
L	mm	Calipers	Measured at several different locations: 73.06, 72.87, 72.87, 72.48, 72.38, 73.07
t	second	Stopwatch	The mass was allowed to fall 10 times, taking these times: 1.95, 1.94, 1.87, 1.85, 1.89, 1.86, 2.05, 1.92, 1.99, 2.02
h	inch	Yardstick	36

Uncertainty Analysis

As mentioned in the "Big Picture", the uncertainty analysis has several aspects:

Transducers—Identify the standard uncertainty for each transducer.
Measurands—For each measurement combine information about the measurements (e.g., several different measurements of the same nominal quantity) with the standard uncertainty of the transducer to determine the standard uncertainty of measurement. Then expand each uncertainty to get the uncertainty of measurement.
Calculated quantity—Propagate the different expanded uncertainties of measurement to determine the uncertainty in the calculated quantity.

We will also consider uncertainty budgets.

Units Note that units of measurement are often an issue for an uncertainty analysis. For this example we chose to convert all measurements and uncertainties to SI units.

Transducer Uncertainty

First we identify the various transducers used during the experiment, and estimate each one's standard uncertainty.

Calipers
All diameters and the length of the rotating drum were measured with digital calipers, with the smallest increment being 0.01 mm. The calipers did not have an in date calibration certificate.

The digital display interval of 0.01 mm of the calipers is *truncated*. Therefore the elemental standard uncertainty due to resolution (see the appendix on resolution) is given by:

$$u_{calipers,resolution} = \frac{0.01}{\sqrt{3}} = 0.005774 \text{ mm} = 5.774 \times 10^{-6} \text{ m}$$

Often when analyzing a process you will find that you do not have a full knowledge of the various uncertainties. That is the case here, where we know that the resolution uncertainty is not the only elemental uncertainty. With no other information about the uncertainty of the calipers, we have to make a decision. We could, for example, assume the uncertainty due to resolution is the only *significant* elemental uncertainty, in which case the final estimate of the standard uncertainty will be an underestimate. An alternative is for us to estimate any additional uncertainty. This might improve the total uncertainty, or it could make it a worse estimate depending upon how good we are at this guesswork! Of course, the best practice would be to have the calipers calibrated, but until we have completed the uncertainty analysis we do not know if calibration will be necessary or cost effective.

For this example we are going to assume that calibration is not a viable option due to either budget or time constraints, and we are going to assume that all other elemental uncertainties combined are about the same as the uncertainty due to resolution. Once we have completed the full uncertainty analysis and developed the uncertainty budgets we can check if this guess is potentially problematic or not. Our guess is thus,

$$u_{calipers,others} = 5.774 \times 10^{-6} \text{ m}$$

We now combine the elemental uncertainties to determine the standard uncertainty of the calipers:

Case Study

$$u_{calipers} = \sqrt{(u_{calipers,resolution})^2 + (u_{calipers,others})^2}$$
$$= \sqrt{(5.774 \times 10^{-6} \text{ m})^2 + (5.774 \times 10^{-6} \text{ m})^2}$$
$$u_{calipers} = 8.165 \times 10^{-6} \text{ m}$$

Balance

The mass m was measured using a calibrated digital balance. The smallest increment on the display was 0.1 g. The calibration certificate gives the total 95% uncertainty of the balance as ±0.13 g.

Since the balance is calibrated, the given uncertainty includes all the elemental uncertainties. Therefore, we do not have to consider resolution—it is already included. If you are using a transducer that has an in date calibration certificate, you will use the uncertainty on the certificate as your transducer uncertainty.

The standard uncertainty for the balance is determined from the 95% uncertainty on the calibration certificate and a k factor of 2.0.

$$u_{balance} = \frac{U}{k} = \frac{0.13}{2.0} = 0.0650 \text{ grams} = 65.0 \times 10^{-6} \text{ kg}$$

Yardstick

The height h is measured with an uncalibrated yardstick. The measurement is in inches, and the smallest divisions on the rule are 1/8 in apart.

As with the uncalibrated calipers, there needs to be some assessment for uncertainty other than just the resolution (for example, nonlinearity, temperature effects, variable scale size). This is another example where you will have to guess. The more information and/or prior knowledge you have about the transducer, the better you can estimate the uncertainty. Also, as with the calipers, you can use the uncertainty budget to check whether your guess is critical to the overall uncertainty or not. If the values you guessed turn out to be significant you will have to address this transducer in more detail, which may include getting it calibrated.

For this exposition we arbitrarily choose to double the resolution uncertainty. Mathematically this calculates to be the same as assuming the increments are ¼-inch apart.

$$u_{yardstick,resolution} = \frac{0.125}{\sqrt{12}} = 0.036084 \text{ in}$$

$$u_{yardstick,total} = 2 \times u_{yardstick,resolution}$$
$$= 2 \times 0.036084$$
$$u_{yardstick,total} = 0.072168 \text{ in}$$

In addition, since the measurement units are inches they have to be converted to m.

$$h(m) = h(\text{inch}) \times 25.4 \times 10^{-3}$$

This is a calculation, so we need to determine the sensitivity by partial differentiation:

$$\frac{\partial h(m)}{\partial h(\text{inch})} = 25.4 \times 10^{-3} \text{ m/inch}$$

The propagated standard uncertainty for the yardstick in meters thus becomes:

$$u_{yardstick} = u_{yardstick,total} \frac{\partial h(m)}{\partial h(\text{inch})} = 0.072168 \times 25.4 \times 10^{-3} = 1.83309 \times 10^{-3} \text{ m}$$

Stopwatch
For this experiment it was decided to measure the drop time manually using an uncalibrated digital stopwatch that truncates time to 0.01 s resolution.

Independently, the operator's human reaction time was tested for this particular test configuration by getting him to time a similar drop which was simultaneously recorded by a very accurate, calibrated high speed camera. With 16 repetitions, the standard deviation of his measurements was 0.0105 s. The average of his measurements was 0.0113 s longer than that obtained from the camera.

Based upon the description, we have four timing uncertainties to include: The *resolution* of the stopwatch; the *variability* due to human reaction; the *systematic delay* caused by human reaction; and "*other*" uncertainties due to, for example, the lack of calibration of the stopwatch.

Resolution
The 0.01 s display resolution is truncated. Therefore the standard uncertainty due to resolution is:

$$u_{stopwatch,resolution} = \frac{0.01}{\sqrt{3}} = 0.005774 \text{ s}$$

Variability
The variability of 0.0105 s is given as a standard deviation based upon 16 observations. From the table, Student-t for $(16-1)=15$ degrees of freedom at 95% confidence is 2.13145. The standard uncertainty due to variability is thus:

$$u_{stopwatch,variability} = \frac{t_{95\%,15}(\text{Standard Deviation})}{2} = \frac{2.13145 \times 0.0105}{2} = 0.0111901 \text{s}$$

Systematic Delay The 0.0113 s difference between the stopwatch and highly accurate measurements can be considered as our best estimate of an error (similar to the

Case Study

error term that can be identified in a calibration process). In this case we will subtract this quantity from each time measurement before using it in any calculations.

While this approach is acceptable when the error has been determined accurately from a complete calibration, the stringent requirements for a calibration were not met in this example, and actual test conditions are unlikely to be the same as those during the experiment to identify human reaction time. While we will still subtract the average delay from each measurement of time since this is our best estimate of the measurement error, we will also include an additional standard uncertainty to take account of uncertainty in this delay.

The systematic delay was determined from the average of 16 observations. Thus, based on the central limit theorem, the average delay is, itself, one number from a normal distribution whose standard deviation is estimated from the sample standard deviation as:

$$(\text{Standard Deviation of the average}) = \frac{(\text{Standard Deviation of the sample})}{\sqrt{(\text{Number of samples})}}$$

Thus:

$$S_{delay} = \frac{0.0105}{\sqrt{16}} = 0.002625\,\text{s}$$

And the standard uncertainty for this component is:

$$u_{stopwatch,delay} = \frac{t_{95\%,15}(\text{Standard Deviation})}{2} = \frac{2.13145 \times 0.002625}{2} = 0.002798\,\text{s}$$

Recall that each time measurement will be reduced by 0.0113 s to account for the average delay.

No Calibration Previously for uncalibrated transducers we chose to account for the lack of calibration by making some assumption about the "other" uncertainties. For this particular stopwatch case we decide that the inherent accuracy of a digital stopwatch is probably much better than the displayed 0.01 s resolution. We also chose to decide that a digital stopwatch is unlikely to lose accuracy, even if it has not been calibrated for some time.

We therefore decide that the uncertainty due to all other aspects is minimal compared with the resolution and human reaction aspects, and chose to ignore them in the uncertainty analysis.

Combined Standard Uncertainty The combined standard uncertainty for the stopwatch is calculated as the RSS of the individual standard uncertainties.

$$u_{stopwatch} = \sqrt{\left(u_{stopwatch,resolution}\right)^2 + \left(u_{stopwatch,variability}\right)^2 + \left(u_{stopwatch,delay}\right)^2}$$

$$= \sqrt{(0.005774)^2 + (0.0111901)^2 + (0.002798)^2}$$

$$u_{stopwatch} = 0.012899\,\text{s}$$

Summary of Transducer Standard Uncertainties

Transducer	Units	Standard uncertainty
Calipers	m	$u_{calipers} = 8.165 \times 10^{-6}$
Balance	kg	$u_{balance} = 65.0 \times 10^{-6}$
Yardstick	m	$u_{yardstick} = 1.83309 \times 10^{-3}$
Stopwatch	second	$u_{stopwatch} = 0.012899$

Uncertainty of Measurement

We now consider each measurand and separately determine its uncertainty of measurement.

Outside Diameter of the Rotating Drum, a

We use the measurements taken at different places around the drum to identify the size and spatial variability of the drum.

When we take several readings for nominally the same measurement, we somehow need to combine those readings to get the "best estimate" of the measurement and its uncertainty. Statistically, the best estimate is the average value, so rather than using any of the individual measurements we use the arithmetic average. We calculate the variability using the central limit theorem[1] which lets us determine the standard deviation we use in the uncertainty analysis (the standard deviation of the mean, $\bar{S}$) calculated from the sample standard deviation, S.

$$\bar{S} = \frac{S}{\sqrt{n}}$$

---

[1] See standard statistics texts for details about the central limit theorem (CLT). For the uncertainty we use here we apply the CLT results that the mean of a random sample drawn from a distribution with standard deviation will have a standard deviation equal to $S/\sqrt{n}$.

Case Study

Number of samples, n	16
Average measurand (m)	52.7606×10^{-3}
Standard deviation (m), S	42.185×10^{-6}
Standard deviation of the mean (m) $\bar{S} = \dfrac{S}{\sqrt{n}}$	$\dfrac{42.185 \times 10^{-6}}{\sqrt{16}} = 10.546 \times 10^{-6}$
Student-t statistic for 95% confidence and $dof = n-1 = 16-1 = 15$	2.13145
Standard uncertainty (m) $u_{spatial} = \dfrac{t_{95\%,dof} \cdot \bar{S}}{2}$	$\dfrac{2.13145 \times 10.546 \times 10^{-6}}{2} = 11.2394 \times 10^{-6}$

We now combine the spatial standard uncertainty and the transducer standard uncertainty to obtain the standard uncertainty of this measurement.

$$u_a = \sqrt{\left(u_{spatial}\right)^2 + \left(u_{calipers}\right)^2}$$

$$u_a = \sqrt{\left(11.2394 \times 10^{-6}\right)^2 + \left(8.165 \times 10^{-6}\right)^2} = 13.892 \times 10^{-6} \text{ m}$$

Finally, determine the expanded uncertainty of measurement for 95% confidence using a k-factor of 2:

$$U_a = k.u_a = 2 \times 13.892 \times 10^{-6} = 27.784 \times 10^{-6} \text{ m}$$

At 95% confidence the average outside diameter of the rotating drum, a, is $52.7606 \times 10^{-3} \pm 27.784 \times 10^{-6}$ m.

Inside Diameter of the Fixed Cylinder, b

This diameter was only measured once, yielding 61.78 mm. Only taking one measurement causes a problem determining the uncertainty for this measurement; there is no deviation for a single number. Faced with this problem, and recognizing that the number is not exact, there can be a number of ways of estimating the uncertainty for this measurement.

We could assume that the only uncertainty is due to the transducer. This approach will lead to an underestimate in the total uncertainty, but may be acceptable for high-tolerance components.

Prior knowledge of the component (or similar components) may offer guidance as to the uncertainty that should be applied to this measurement.

One effective approach is to assume a "reasonable" number for this uncertainty. We then use an uncertainty budget to identify whether this measurement is critical to the final uncertainty or not. If the budget shows this uncertainty is significant then this uncertainty component needs more attention. This is the approach adopted for this example.

For this example, let us assume that you have given the fixed cylinder and the rotating drum a cursory examination and decided that the fixed cylinder is manufactured to the same or better tolerance than the rotating drum. Therefore, we decide to assume that the standard uncertainty due to spatial variation in the diameter of the cylinder is the same as the standard uncertainty for the rotating drum that we calculated earlier as:

$$u_{b,spatial} = 11.2394 \times 10^{-6} \, \text{m}$$

As a consequence, since both diameters were measured with the same transducer (calipers), their combined standard uncertainty and also their expanded uncertainty of measurement will be the same.

$$u_b = \sqrt{(u_{b,spatial})^2 + (u_{calipers})^2}$$

$$u_b = \sqrt{(11.2394 \times 10^{-6})^2 + (8.165 \times 10^{-6})^2} = 13.892 \times 10^{-6} \, \text{m}$$

$$U_b = k.u_b = 2 \times 13.892 \times 10^{-6} = 27.784 \times 10^{-6} \, \text{m}$$

At 95% confidence the inside diameter of the fixed cylinder, b, is $61.78 \times 10^{-3} \pm 27.784 \times 10^{-6}$ m.

Diameter of the Pulley Fastened to the Top of the Rotating Drum, d

Similar to the measurement of the inside diameter of the fixed cylinder, b, there was only one measurement of the diameter of this pulley; $d = 25.4 \, \text{mm}$. Let us assume a perfunctory inspection identified that this pulley was made with "some degree of accuracy". However, the effective radius at which the tension is applied is not well defined because of the unknown thickness of the string wrapped around the pulley. We will arbitrarily decide to include an uncertainty (estimated at 95% confidence) of 0.5 mm to take this into account, and leave it to the uncertainty budget to identify whether we need to improve on this measurement or not. Of course, our 0.5 mm estimate may be hopelessly wrong, but it is the best we can do at this stage without additional information or measurements!

$$u_{d,other} = \frac{0.5 \times 10^{-3}}{k} = \frac{0.5 \times 10^{-3}}{2} = 0.25 \times 10^{-3} \, \text{m}$$

We now combine the "other" standard uncertainty and the transducer standard uncertainty to obtain the standard uncertainty of this measurement.

$$u_d = \sqrt{(u_{d,other})^2 + (u_{calipers})^2}$$

$$u_d = \sqrt{(0.25 \times 10^{-3})^2 + (8.165 \times 10^{-6})^2} = 250.13 \times 10^{-6} \, \text{m}$$

Case Study

Finally, determine the expanded uncertainty of measurement for 95% confidence using a k-factor of 2:

$$U_d = k.u_d = 2 \times 250.13 \times 10^{-6} = 500.3 \times 10^{-6}\, m$$

At 95% confidence the diameter of the pulley, k, is $25.4 \times 10^{-3} \pm 500 \times 10^{-6}\, m$.

Mass, m

The length measurements had spatial variability due to irregularity in the actual size of the components. In comparison, the mass only has one true value. The measurement is the best estimate we have of that mass, and its uncertainty is the uncertainty from the balance's calibration certificate.

Therefore:

$$u_{mass} = u_{balance} = 65.0 \times 10^{-6}\, kg$$

And the expanded uncertainty of the measurement is

$$U_{mass} = k.u_{mass} = 2 \times 65.0 \times 10^{-6} = 130 \times 10^{-6}\, kg$$

At 95% confidence the mass, m, is $100.3 \times 10^{-3} \pm 130 \times 10^{-6}\, kg$.

Length of the Rotating Drum (from Fluid Surface to Bottom of Drum), L

We use the same analysis as was used for the outside diameter of the rotating drum. Only the numbers have been changed!

Number of samples, n	6
Average measurand (m)	72.78833×10^{-3}
Standard deviation (m), S	292.672×10^{-6}
Standard deviation of the mean (m) $\bar{S} = \dfrac{S}{\sqrt{n}}$	$\dfrac{292.672 \times 10^{-6}}{\sqrt{6}} = 119.4827 \times 10^{-6}$
Student-t statistic for 95% confidence and $dof = n-1 = 15$	2.570582
Standard uncertainty (m) $u_{spatial} = \dfrac{t_{95\%,dof} \cdot \bar{S}}{2}$	$\dfrac{2.570582 \times 119.4827 \times 10^{-6}}{2} = 153.5700 \times 10^{-6}$

We now combine the spatial standard uncertainty and the transducer standard uncertainty to obtain the standard uncertainty of this measurement.

$$u_L = \sqrt{(u_{spatial})^2 + (u_{calipers})^2}$$

$$u_L = \sqrt{(153.5700\times10^{-6})^2 + (8.165\times10^{-6})^2} = 153.787\times10^{-6}\,\text{m}$$

Finally, determine the expanded uncertainty of measurement for 95% confidence using a *k*-factor of 2:

$$U_L = k \cdot u_L = 2\times153.787\times10^{-6} = 307.574\times10^{-6}\,\text{m}$$

At 95% confidence the average length of the rotating drum, L, is $72.78833\times10^{-3} \pm 307.574\times10^{-6}$ m.

Time it Takes the Mass to Fall, t

We use a similar analysis as was used for the outside diameter and the length of the rotating drum. For this measurand, though, we need to include the delay aspect of the human reaction time.

Number of samples, n	10
Average measurand (s)	1.9340
Average delay (from stopwatch calculations) (s)	0.0113
Average value less delay (s)	$1.934 - 0.0113 = 1.9227$
Standard deviation (s), S	0.06915
Standard deviation of the mean(s) $\bar{S} = \dfrac{S}{\sqrt{n}}$	$\dfrac{0.06915}{\sqrt{10}} = 0.021868$
Student-t statistic for 95% confidence and $dof = n-1 = 9$	2.262157
Standard uncertainty (s) $u_{time} = \dfrac{t_{95\%,dof}\cdot\bar{S}}{2}$	$\dfrac{2.262157\times 0.021868}{2} = 0.024735$

We now combine the standard uncertainty of the time measurements with the standard uncertainty of the stopwatch, determined previously as $u_{stopwatch} = 0.012899$ s.

$$u_t = \sqrt{(u_{time})^2 + (u_{stopwatch})^2}$$

$$u_t = \sqrt{(0.024735)^2 + (0.012899)^2} = 0.02790\,\text{s}$$

Finally, determine the expanded uncertainty of measurement for 95% confidence using a k-factor of 2:

$$U_t = k \cdot u_t = 2 \times 0.02790 = 0.05579 \text{s}$$

At 95% confidence the average drop time, t, is 1.9227 ± 0.05579 s.

Caution: Have we inadvertently included human reaction time twice? This was originally assessed when the uncertainty of the stopwatch was estimated, and it seems it might also have been included in the above calculation.

This type of issue is one that can crop up from time to time. In this particular instance, the variability in time measurement is a combination of human reaction time and inherent variability in the actual time due to variations from test to test. There is thus some double-accounting. However, with the given information it is not obvious how to extract the different components, especially since the test conditions are probably variable. In this particular case we decide to leave it to the uncertainty budget to identify whether this is a critical measurement or not.

Distance Mass Falls at Terminal Velocity, h

The distance of 36 in is converted to 0.9144 m. Similar to the measurement of the mass, the measurement of drop distance is a single measurement of a quantity that has no spatial variation. The measurement uncertainty is the uncertainty of the yardstick. We calculate the expanded uncertainty from the standard uncertainty:

$$U_{distance} = k \cdot u_{yardstick} = 2 \times 1.8331 \times 10^{-3} = 3.6662 \times 10^{-3} \text{ m}$$

At 95% confidence the drop distance, h, is $0.9144 \text{ m} \pm 3.6662 \times 10^{-3}$ m.

Acceleration Due to Gravity, g

Very often the acceleration due to gravity is assumed to be constant, and standard numbers are used in calculations. These values may be detailed for a particular industry, company or test specification, in which case the prescribed number should be used. Sometimes the value is left to the analyst. Especially for tests that are "gravity sensitive" a full uncertainty analysis should include variability in the final answer due to uncertainty in the assumed acceleration due to gravity to verify that the acceleration due to gravity is not a critical quantity.

> When using SI units, some typical assumed values are: 9.8, 9.8065, 9.81 and 10 m/s².
> When using US customary units, some typical assumed values are: 32, 32.17 and 32.2 ft/s².
> When the unit basis is inches (ips), typical assumed values are: 386.4 and 386 in/s².

The actual acceleration due to gravity at a test facility depends upon a number of factors, which can include local geological features such as the density of bedrock. The nominal acceleration due to gravity at a test facility can calculated using the International Gravity Formula (IGF) of 1980 and the latest World Geodetic System cartographic model WGS84[2], which take into account the rotation of the Earth, the height above sea level, and the oblate spheroidal shape of the geoid.

$$g = 9.7803267714 \left(\frac{1+0.00193185138639(sin(\phi))^2}{\sqrt{1-0.00669437999013(sin(\phi))^2}} \right) \left(\frac{R}{R+e} \right)^2$$

Where ϕ is the latitude and e is the elevation above sea level of the test facility, and $R = 6,371,000$ m is the nominal radius of the earth.

We will assume that the experiment described here was conducted at a location that is 60±45 m above sea level, at a latitude of 38°50'±50'. This is equivalent to the rather coarse measurement of saying that the test facility is somewhere in Maryland. A separate uncertainty analysis of the IGF for this location gives (95% confidence assumed):

$$g = 9.8007868 \pm 749.05 \times 10^{-3} \, \text{m/s}^2$$

For this ongoing uncertainty analysis we will therefore elect to use $g = 9.8007$ m/s², with a 95% uncertainty of $U_g = 0.000749$ m/s².

> At 95% confidence the acceleration due to gravity, g, is 9.8007 ± 0.00075 m/s².

---

[2] WGS84 is the coordinate system used by Global Positioning System, GPS.

Case Study

Propagation of Uncertainty

The measured values and their 95% uncertainties are summarized. Recall that the actual calculations were conducted in Excel, carrying forward full significance. The numbers here are rounded:

Quantity	Value	Uncertainty, U (95% confidence)	Units of measurement
a	52.76×10^{-3}	27.78×10^{-6}	m
b	61.78×10^{-3}	27.78×10^{-6}	m
d	25.4×10^{-3}	500.3×10^{-6}	m
m	100.3×10^{-3}	130×10^{-6}	kg
L	72.79×10^{-3}	307.6×10^{-6}	m
t	1.9227	55.79×10^{-3}	s
h	0.9144	3.66×10^{-3}	m
g	9.8007	0.75×10^{-3}	m/s^2

The equation that determines the *calculated* quantity (the fluid's viscosity) is the following function of the *measured* quantities.

$$\mu = \left[\frac{mgd^2 (b-a)}{2\pi L a^2 b} \right] \frac{t}{h}$$

From which the sensitivities of the viscosity with respect to each measured quantity are determined from the partial differential equations as follows:

Quantity	Sensitivity	Evaluated	Units of sensitivity
a	$\dfrac{gd^2 mt(a-2b)}{2\pi L a^3 bh}$	-22.751	$(\text{N.s/m}^2)/\text{m}$
b	$\dfrac{gd^2 mt}{2\pi Lab^2 h}$	14.479	$(\text{N.s/m}^2)/\text{m}$
d	$\dfrac{gdmt(b-a)}{\pi L a^2 bh}$	12.041	$(\text{N.s/m}^2)/\text{m}$
m	$\dfrac{gd^2 t(b-a)}{2\pi L a^2 bh}$	1.5246	$(\text{N.s/m}^2)/\text{kg}$
L	$\dfrac{gd^2 mt(a-b)}{2\pi L^2 a^2 bh}$	-2.1009	$(\text{N.s/m}^2)/\text{m}$
t	$\dfrac{gd^2 m(b-a)}{2\pi L a^2 bh}$	0.07953	$(\text{N.s/m}^2)/\text{s}$
h	$\dfrac{gd^2 mt(a-b)}{2\pi L a^2 bh^2}$	-0.16723	$(\text{N.s/m}^2)/\text{s}$
g	$\dfrac{d^2 mt(b-a)}{2\pi L a^2 bh}$	0.015603	$(\text{N.s/m}^2)/(\text{m/s}^2)$

We now multiply each uncertainty of measurement by its related sensitivity. This determines a quantity that:
- Has the same engineering units as the calculated quantity.
- Is the "contribution" of uncertainty due to that measurement.

Quantity	Uncertainty of measurement, U from earlier table (95% confidence)	Sensitivity from previous table	Contribution (Uncertainty) × (Sensitivity)
a	27.78×10^{-6}	-22.751	-0.6321×10^{-3}
b	27.78×10^{-6}	14.479	0.4023×10^{-3}
d	500.3×10^{-6}	12.041	6.024×10^{-3}
m	130×10^{-6}	1.5246	0.1982×10^{-3}
L	307.6×10^{-6}	-2.1009	-0.6462×10^{-3}
t	55.79×10^{-3}	0.07953	4.4373×10^{-3}
h	3.66×10^{-3}	-0.16723	-0.6131×10^{-3}
g	0.75×10^{-3}	0.015603	0.01169×10^{-3}

The final uncertainty of measurement is the root sum of the squares (RSS) of the contributions:

$$\text{Uncertainty of viscosity}, U_\mu = 0.007574 \text{ N.s/m}^2$$

Calculating the viscosity:

$$\mu = \left[\frac{mgd^2(b-a)}{2\pi La^2 b}\right]\frac{t}{h} = 0.15292 \text{ N.s/m}^2$$

Bringing together the final calculation and its uncertainty, with appropriate rounding we can state that:

With 95% confidence the viscosity of the fluid under test was $0.1529 \pm 0.0076 \text{ N.s/m}^2$ Or, at 95% confidence the true viscosity is in the range $0.1453 < \mu < 0.1605 \text{ N.s/m}^2$

Case Study   79

We now reveal that the fluid under test was SAE 10W-30 oil. The average temperature of the oil during testing was 19.2 °C. Reference graphs for SAE 10W-30 give, for this temperature, a viscosity of 0.148 N.s/m². Fortunately the uncertainty range for the experiment includes the tabulated value!

We see that the uncertainty represents ±5% of the calculated value. Assuming that this is deemed unacceptably high we now turn to *uncertainty budgets* to help improve the situation.

Uncertainty Budgets

The GUM gives very little guidance on uncertainty budgets, and yet they are arguably one of the most useful products of an uncertainty analysis. In addition to the uncertainty values, different budgets can factor in a wide variety of aspects. For example, the cost of equipment can be included (see the worked example in Chap. 5 concerning a heat transfer measurement).

Also, there are different levels of uncertainty budget that can be developed. A "low-level[3]" budget may include relatively little information and be primarily concerned with measurement uncertainty. A "high-level" uncertainty budget can include much more information, and can identify specific problematic elemental uncertainties.

In this exposition we first introduce a low-level budget (one with a low level of detail). We then expand the budget to include more detail. Finally we will give a comprehensive high-level budget. None of the budgets presented here include any new numbers—they just bring together some of the uncertainty data calculated previously in an easy-to-read format.

A Low-Level Uncertainty Budget
This uncertainty budget includes little more than the different measurement uncertainties and their contribution to the propagated uncertainty in the measured quantity. All of the numbers have already been calculated above, and are repeated in the uncertainty budget table.

---

[3] The terms high-level, middle-level and low-level are not included in the GUM. They are used here purely to give an indication of the range of content that can be presented in different uncertainty budgets.

Calculated Viscosity: 0.1529 N.s/m²
95% uncertainty: ±0.0076 N.s/m²

Measurement	Symbol	95% uncertainty	Propagated uncertainty, or contribution $(N.s/m^2)$
Outside diameter of the rotating drum	a	27.78×10^{-6}	-0.6321×10^{-3}
Inside diameter of the fixed cylinder	b	27.78×10^{-6}	0.4023×10^{-3}
Diameter of the pulley fastened to the top of the rotating drum	d	500.3×10^{-6}	6.024×10^{-3}
Mass suspended on the string	m	130×10^{-6}	0.1982×10^{-3}
Length of the rotating drum (from fluid surface to bottom of drum)	L	307.6×10^{-6}	-0.6462×10^{-3}
Time it takes the mass to fall distance h	t	55.79×10^{-3}	4.4373×10^{-3}
Distance mass falls at terminal velocity	h	3.66×10^{-3}	-0.6131×10^{-3}
Acceleration due to gravity	g	0.75×10^{-3}	0.01169×10^{-3}

There are several observations we can make:

If the uncertainty in viscosity of ±0.0076 N.s/m² is adequate for the purpose, then no changes need to be made to the apparatus or test protocol. Thus, if the reason you are checking a sample is to see if it is oil, water or honey, the exercise is complete.

Also, if the uncertainty is acceptable you have demonstrated that it is not necessary to calibrate the calipers, yardstick or stopwatch.

For an engineering test this experiment probably has an inadequately high amount of uncertainty. We can use the uncertainty analysis and budget as a tool to help justify changing the experiment.

The contribution due to uncertainty in acceleration due to gravity is the smallest by more than an order of magnitude. Thus the significant lack of knowledge of latitude or elevation above sea level (we stated the test facility was "somewhere in Maryland") has not made any significant difference to the results. As far as uncertainty is concerned, it would be a waste of time and resources identifying the exact location of the test facility in order just to improve the estimate of the acceleration due to gravity.

Several of the parameters have comparable propagated contributions in the range of about 0.2 to 0.6×10^{-3} N.s/m². The uncertainty in each of these parameters contributes approximately equally to the final uncertainty. There is likely to be little return on investment if just one of these parameters is assessed for improvement.

The two largest contributions are those due to the uncertainties in the diameter of the pulley on top of the apparatus, and the time it takes the mass to fall. If the overall uncertainty needs to be reduced, these are the parameters that have to be measured more accurately.

A Medium-Level Uncertainty Budget

The next level of uncertainty budget includes the same information that we saw in the low-level budget. However, this time we decide to separate the uncertainty due

Case Study

to the transducers from the uncertainty due to variability in the quantities being measured. As before, most of the numbers have already been given during the analysis, and are repeated in the uncertainty budget table. New numbers for the table are standard uncertainties, u, that have been expanded to 95% confidence uncertainties, U, using a k-factor of 2 and:

$$U = k.u = 2.u$$

Calculated Viscosity: 0.1529N.s/m^2
95% uncertainty: $\pm 0.0076 \text{N.s/m}^2$

Measurement, symbol and units		95% uncertainty	Measurement 95% uncertainty	Propagated uncertainty, or contribution (N.s/m^2)
Outside diameter of the rotating drum, a (m)	Calipers	16.33×10^{-6}	27.78×10^{-6}	-0.621×10^{-3}
	Measurand	22.5×10^{-6}		
Inside diameter of the fixed cylinder, b (m)	Calipers	16.33×10^{-6}	27.78×10^{-6}	0.4023×10^{-3}
	Measurand	22.5×10^{-6}		
Diameter of the pulley fastened to the top of the rotating drum, d (m)	Calipers	16.33×10^{-6}	500.3×10^{-6}	6.024×10^{-3}
	Measurand	500×10^{-6}		
Mass suspended on the string, m (kg)	Balance	130.0×10^{-6}	130×10^{-6}	0.1982×10^{-3}
	Measurand	Zero		
Length of the rotating drum (from fluid surface to bottom of drum), L (m)	Calipers	16.33×10^{-6}	307.6×10^{-6}	-0.6462×10^{-3}
	Measurand	307×10^{-6}		
Time it takes the mass to fall, t (s)	Stopwatch	25.80×10^{-3}	55.79×10^{-3}	4.4373×10^{-3}
	Measurand	49.47×10^{-3}		
Distance mass falls at terminal velocity, h (m)	Yardstick	3.67×10^{-3}	3.666×10^{-3}	-0.6131×10^{-3}
	Measurand	Zero		
Acceleration due to gravity, $g\left(\text{m/s}^2\right)$	IGF, Latitude	736×10^{-6}	749×10^{-6}	0.01169×10^{-3}
	IGF, Elevation	138×10^{-6}		

Entries marked "zero" indicate a single measurement, and therefore no assessment of (e.g., spatial) variation can be made

Since this medium-level budget includes the same information as was in the low-level budget, clearly we can make the same observations as before. However, the additional information in this budget also helps us focus our interest. Here are some more observations we could make:

> For each measurement that used calipers the uncertainty due to the transducer was significantly less than the uncertainty due to the measurand variability. Therefore there is a good argument that the calipers do not need calibration.

> The main source of uncertainty is from the measurement of the pulley on top of the rotating drum. This medium-level uncertainty budget has identified that virtually all of this uncertainty is due to the measurand itself, and virtually none is due to uncertainty in the transducer. Therefore, putting effort into getting a better estimate (measurement) of the effective diameter of the pulley should be the primary focus for improving this experiment, but using the calipers is acceptable.

> For the time measurements the uncertainty due to the transducer (stopwatch) is about half of that due to measurand variability. Therefore the stop watch probably does not need calibrating. Improvements can be made by changing the timing method (for example, having an automated system that eliminates the human reaction time aspect). However, for this measurand there is still a miscellany of experimental inputs that we have not identified or separately accounted for.

High-Level Uncertainty Budget

We now look at an example of a high-level uncertainty budget. Essentially the table for this budget includes all of the numbers generated earlier in this example. We choose to display all elemental uncertainties as 95% confidence expanded elemental uncertainties rather than as standard uncertainties so that we can directly compare all uncertainties in the table; thus where the previous work shows standard uncertainties, we modify them before entering them in this table:

$$95\% \text{ confidence expanded elemental uncertainty,}$$
$$U = k.(\text{elemental standard uncertainty}, u)$$

$$U = 2.u$$

Also, to fully populate the table we need to know the propagated uncertainty contribution due to each elemental uncertainty. Strictly we should determine each of these from separate uncertainty propagations—a lot of work! We can, though, "cheat" and can determine the individual contributions from what is essentially a reverse calculation. As we suggested earlier, the uncertainty analysis was determined using a spreadsheet. We also recognize that all uncertainties are combined using RSS (root sum of squares) calculations. Let us consider an example where we only have three individual uncertainties: U_1, U_2 and U_3. We wish to identify the contribution due to one of them, U_1, when we already know the total uncertainty, U_{total}. In the spreadsheet we temporarily set U_1 to zero and let the spreadsheet calculate the modified total uncertainty, $U_{temporary}$.

U_1 Unknown contribution due to elemental uncertainty 1
U_{total} Known total uncertainty
$U_{temporary}$ Spreadsheet total uncertainty with U_1 set to zero

From the uncertainty analysis RSS calculation we have:

$$U_{total}^2 = U_1^2 + U_2^2 + U_3^2$$

And from the spreadsheet with U_1 set to zero:

$$U_{temporay}^2 = U_2^2 + U_3^2$$

Combine and rearrange these equations to determine U_1:

$$U_1 = \sqrt{U_{total}^2 - U_{temporary}^2}$$

From the full analysis we already know the total uncertainty in the calculated quantity, U_{total}. Separately for each elemental uncertainty we temporarily delete the entry in the spreadsheet, and identify the new uncertainty in the calculated quantity, $U_{temporary}$. The propagated contribution due to the deleted elemental uncertainty can now be calculated as:

$$(\text{propagated elemental contribution}) = \sqrt{U_{total}^2 - U_{temporary}^2}$$

If you are using a spreadsheet that already holds the complete uncertainty analysis, this is a fairly quick and easy calculation that can be used to determine the contribution for elemental uncertainties without having to conduct a complete uncertainty propagation for every individual component.

In the table below we have also included a column labeled "Rank." This sorts the contributions from first (largest contribution) in descending order to twenty-first (least problematic and smallest contribution). As we have seen previously, you have a lot of choice over what information you can include in an uncertainty budget. The "Rank" column is another example of something nice to have, but that is not essential.

Case Study

Calculated Viscosity: $0.1529 \, \text{N.s/m}^2$

95% uncertainty: $\pm 0.0076 \, \text{N.s/m}^2$

Measurement, symbol (units of measure)	Source of uncertainty	95% expanded uncertainty (units of measure)	Contribution (N.s/m^2)	Rank
Outside diameter of the rotating drum, a (m)	Calipers, resolution	11.55×10^{-6}	262.7×10^{-6}	11=
	Calipers, other	11.55×10^{-6}	262.7×10^{-6}	11=
	Spatial variability	22.48×10^{-6}	606.2×10^{-6}	6
Inside diameter of the fixed cylinder, b (m)	Calipers, resolution	11.55×10^{-6}	167.2×10^{-6}	14=
	Calipers, other	11.55×10^{-6}	167.2×10^{-6}	14=
	Spatial variability	22.48×10^{-6}	325.5×10^{-6}	10
Diameter of the pulley fastened to the top of the rotating drum, d (m)	Calipers, resolution	11.55×10^{-6}	139.0×10^{-6}	16=
	Calipers, other	11.55×10^{-6}	139.0×10^{-6}	16=
	Diameter, other	0.500×10^{-3}	6020.4×10^{-6}	1
Mass suspended on the string, m (kg)	Balance calibration	130.0×10^{-6}	198.2×10^{-6}	13
Length of the rotating drum (from fluid surface to bottom of drum), L (m)	Calipers, resolution	11.55×10^{-6}	24.3×10^{-6}	18=
	Calipers, other	11.55×10^{-6}	24.3×10^{-6}	18=
	Spatial variability	307.1×10^{-6}	645.3×10^{-6}	5
Time it takes the mass to fall, t (s)	Stopwatch, resolution	11.547×10^{-3}	918.4×10^{-6}	4
	Stopwatch, variability	22.380×10^{-3}	1780.0×10^{-6}	3
	Stopwatch, delay	5.595×10^{-3}	445.0×10^{-6}	9
	Time variability	49.470×10^{-3}	3934.5×10^{-6}	2
Distance mass falls at terminal velocity, h (m)	Yardstick, resolution	1.833×10^{-3}	531.0×10^{-6}	7=
	Yardstick, other	1.833×10^{-3}	531.0×10^{-6}	7=
Acceleration due to gravity, $g \, (\text{m/s}^2)$	Latitude (radians)	14.544×10^{-3}	11.5×10^{-6}	20
	Elevation above sea level (m)	45	2.2×10^{-6}	21

So has all the extra effort needed to produce the high-level uncertainty budget given us any more useful information?

> We saw from the previous budgets that uncertainty in the acceleration due to gravity was the smallest contributor to the uncertainty in viscosity. This high-level budget confirms that the two gravity elemental uncertainties (latitude and elevation above sea level, ranked twentieth and twenty-first) are the two smallest uncertainty contributions in the table.
>
> The calipers had a resolution (smallest scale increment) of 0.01 mm. Looking at the different rows in the budget where the caliper resolution appears, we see that this elemental uncertainty is always ranked low on this list of problems. Imagine your boss tells you he wants the viscosity measured more accurately, and to do this he wants you to buy calipers that measure to 0.001 mm and have them calibrated. Here's your chance to show him why he would be wasting his money (don't forget to ask for a bonus as well!).
>
> The fact that the uncertainty in pulley diameter ranked first is probably no surprise. The pulley is fairly small (25.4 mm diameter), and the string introduced an estimated uncertainty of about 0.5 mm to that value. Clearly this is an area where the experiment needs attention, and the uncertainty budget has confirmed that poor knowledge of the effective diameter is a big problem. However, the calipers we used to measure the pulley diameter are OK! So maybe you need to use some of that bonus to buy thinner string and/or redesign the pulley to reduce that uncertainty? Maybe double the diameter of the pulley and half the size of the mass, or something along those lines?
>
> Since the drop time was measured manually with a stopwatch, it is probably no surprise that the stopwatch uncertainties rank fairly high (third, fourth and ninth). Thus, after you have fixed the pulley problem it might be worth investing more of your bonus in a more accurate method of measuring the time.
>
> What is probably more surprising is that the actual time measurements also have significant time variability, ranking second. You also need to investigate why the drop time is varying so much. This might be a human reaction timing problem, or it might be something going on in the experiment itself. The uncertainty budget can't resolve that question, but it did lead you to the problem area!

Uncertainty Budget Summary

An uncertainty budget is essentially a table that brings together information about the different uncertainties in a process. The table can also include other information. For example, we saw an example in Chap. 5 concerning heat transfer from a hot plate to a cold plate that costs of different equipment could also be included in the budget.

The uncertainty budget rarely answers questions, but it does guide you to where there are problems that need addressing. Of equal importance, it can help identify aspects of a process where committing additional resources is not warranted.

Appendix

This appendix develops the mathematics of resolution, scale size, truncation and rounding. While the development may be educational in its own right, the approach can be modified to deal with other uncertainties that have industry-specific statistical distributions.

The Mathematics of Resolution and Truncation

Imagine that you are using an analog pressure gage that you can read to the nearest 10 psi. How do you calculate the standard uncertainty for that 10 psi gage increment? Is the standard uncertainty the same if you are using a digital display that truncates the readings, rather than rounds them?

These are the types of problems we address in this appendix. We do this by assuming that the distribution of true values along the number line representing the measurement is a uniform distribution. This is probably a good distribution to use. The normal distribution would not be a good fit here because there is no reason why the true values should "clump up" around the divisions of the measurement transducer.

We will look at the mathematics of three different scenarios and develop the equations for converting scale increment to standard uncertainty, the building block to uncertainty analysis that was presented in the main body of this booklet. The scenarios we investigate are:

Calculating the standard uncertainty from the increment size (full resolution)
Calculating the standard uncertainty from half of the increment size (half resolution)
Calculating the standard uncertainty for digital displays that truncate to the least significant digit

Summary of the results from this appendix:

If a_{FULL} is the **FULL-RESOLUTION**, e.g., if the smallest division on a rule is 1 mm the full-resolution is $a_{FULL} = 1\text{mm}$, and the measurement is **ROUNDED** to the nearest division, the standard uncertainty is given by:

$$u_{RESOLUTION} = \frac{a_{FULL}}{\sqrt{12}}$$

If a_{HALF} is the **HALF-RESOLUTION**, e.g., if the smallest division on a rule is 1 mm the half-resolution is $a_{HALF} = 1/2 = 0.5\text{mm}$, and the measurement is **ROUNDED** to the nearest division, the standard uncertainty is given by:

$$u_{RESOLUTION} = \frac{a_{HALF}}{\sqrt{3}}$$

When a_{SCALE} is the smallest increment on a digital display, for example the last digit on a display reads to 0.01 in then $a_{SCALE} = 0.01\text{in}$, and the readings are **TRUNCATED** to the next lowest division the standard uncertainty is:

$$u_{TRUNCATED} = \frac{a_{SCALE}}{\sqrt{3}}$$

When a_{SCALE} is the smallest increment on a digital display, for example the last digit on a display reads to 0.01 in then $a_{SCALE} = 0.01\text{in}$, and the readings are **TRUNCATED AND ADJUSTED FOR SYSTEMATIC ERROR**, the standard uncertainty is:

$$u_{ADJUSTED} = \frac{a_{SCALE}}{\sqrt{12}}$$

When using an **ANALOG-TO-DIGITAL CONVERTER** with *n* bits set to a maximum input of V_{MAX}:

$$\text{with rounding, Standard uncertainty} = \frac{V_{MAX}}{2^n \sqrt{12}}$$

$$\text{with truncation, Standard uncertainty} = \frac{V_{MAX}}{2^n \sqrt{3}}$$

Remember that even though this appendix uses statistics to determine the standard uncertainties (which suggests that resolution is a Type A uncertainty), the GUM specifically mentions resolution as Type B, and you should assess it as such.

Appendix

Derivation of the Standard Uncertainty When Using the Full-Resolution

If an analog pressure gage has increments every 10 psi, and the pressure is only being recorded at those increment levels (no guessing at readings between the increments), and you round the pressure to the nearest increment, the *full-resolution* of the gage is 10 psi.

Let the full-resolution (smallest scale size) be a_{FULL}, in which case the probability density function (pdf) for the uniform distribution for a measurement reading h is given by:

$$f(x) = \begin{cases} \dfrac{1}{a_{FULL}} & \text{for } \left(h - \dfrac{a_{FULL}}{2}\right) \leq x \leq \left(h + \dfrac{a_{FULL}}{2}\right) \\ 0 & \text{elsewhere} \end{cases}$$

Note that this function satisfies all the requirements of a pdf, one being that $\int_{-\infty}^{+\infty} f(x)\,dx = 1$ since

$$\int_{-\infty}^{+\infty} f(x)\,dx = \int_{h - \frac{a_{FULL}}{2}}^{h + \frac{a_{FULL}}{2}} \frac{1}{a_{FULL}}\,dx$$

$$= \left[\frac{x}{a_{FULL}}\right]_{h - \frac{a_{FULL}}{2}}^{h + \frac{a_{FULL}}{2}} = \left(\frac{h}{a_{FULL}} + \frac{a_{FULL}}{2a_{FULL}} - \frac{h}{a_{FULL}} + \frac{a_{FULL}}{2a_{FULL}}\right) = 1$$

The variance of the distribution is given by:

$$var = \int_{-\infty}^{+\infty} (x-h)^2 f(x)\,dx = \int_{h - \frac{a_{FULL}}{2}}^{h + \frac{a_{FULL}}{2}} (x-h)^2 \frac{1}{a_{FULL}}\,dx$$

$$= \left[\frac{(x-h)^3}{3a_{FULL}}\right]_{h - \frac{a_{FULL}}{2}}^{h + \frac{a_{FULL}}{2}} = \left(\left(h + \frac{a_{FULL}}{2} - h\right)^3 - \left(h - \frac{a_{FULL}}{2} - h\right)^3\right)\frac{1}{3a_{FULL}} = \frac{a_{FULL}^2}{12}$$

For this uniform distribution, the standard uncertainty is numerically the same as the standard deviation, so:

$$u_{RESOLUTION} = \text{standard deviation} = \sqrt{var} = \frac{a_{FULL}}{\sqrt{12}}$$

REMEMBER, a_{FULL} is the FULL-RESOLUTION, and readings are recorded to the nearest division on the scale. For example, if the smallest divisions on a ruler are in 1 mm intervals, then the full-resolution $a_{FULL} = 1\text{mm}$ and the standard uncertainty is

$$u_{RESOLUTION} = \frac{a_{FULL}}{\sqrt{12}} = \frac{1}{\sqrt{12}} \approx 0.289\,\text{mm}$$

Derivation of the Standard Uncertainty When Using the Half-Resolution

If an analog pressure gage has increments every 10 psi, and the pressure is only being recorded at those increment levels (still no guessing at readings between the increments), the *full-resolution* of the gage is 10 psi.

However, some people argue that the true value should never be more than half of the scale increment from the reading (i.e., 10/2=5 psi). For example, if the true pressure is 196 psi, it will be recorded as 200 psi, (4 psi too high). A true pressure of 194 psi will be recorded as 190 psi (4 psi too low). In this case those folk state that the resolution of the 10 psi increment pressure gage is 5 psi. In this appendix we refer to this resolution as the *half-resolution*, a_{HALF}.

$$a_{HALF} = \frac{a_{FULL}}{2} = \frac{(\text{scale increment})}{2}$$

The probability density function for the uniform distribution for a measurement reading h is given by:

$$f(x) = \begin{cases} \dfrac{1}{2a_{HALF}} & \text{for } (h - a_{HALF}) \leq x \leq (h + a_{HALF}) \\ 0 & \text{elsewhere} \end{cases}$$

The variance of the distribution is given by:

$$\text{var} = \int_{-\infty}^{+\infty} (x-h)^2 f(x)\,dx = \int_{h-a_{HALF}}^{h+a_{HALF}} (x-h)^2 \frac{1}{2a_{HALF}}\,dx = \left[\frac{(x-h)^3}{6a_{HALF}}\right]_{h-a_{HALF}}^{h+a_{HALF}}$$

$$= \left((h + a_{HALF} - h)^3 - (h - a_{HALF} - h)^3\right)\frac{1}{6a_{HALF}} = \frac{a_{HALF}^2}{3}$$

Appendix

For this uniform distribution, the standard uncertainty is again numerically the same as the standard deviation:

$$u_{RESOLUTION} = \text{standard deviation} = \sqrt{var} = \frac{a_{HALF}}{\sqrt{3}}$$

REMEMBER, a_{HALF} is the HALF-RESOLUTION. For example, if the smallest divisions on a ruler are in 1 mm intervals, then the half-resolution $a_{HALF} = 0.05$ mm and the standard uncertainty is $u_{RESOLUTION} = \frac{a_{HALF}}{\sqrt{3}} = \frac{0.5}{\sqrt{3}} \approx 0.289$ mm, which is the same result determined for this ruler previously when considering the full-resolution. The standard uncertainty doesn't depend upon your definitions. It depends upon the equipment you are using!

Derivation of the Standard Uncertainty When Values Are Truncated

The previous examples assumed that the distribution of readings was centered about the true value. This is because the readings were rounded to the nearest scale division. For truncation we have a different situation. For example, a true pressure of 196 psi will truncate to 190 psi, but round to 200 psi. Similarly, a true pressure of 199.9 psi also truncates to 190 psi.

The consequence is that the expected value is half of the scale resolution larger than the scale increment—the distribution is one sided rather than symmetric.

Let the scale size be a_{SCALE}, in which case the probability density function for the uniform distribution for a measurement reading h is given by:

$$f(x) = \begin{cases} \dfrac{1}{a_{SCALE}} & \text{for}(h) \le x \le (h + a_{SCALE}) \\ 0 & \text{elsewhere} \end{cases}$$

The average, $\bar{x}$, is given by:

$$\bar{x} = \int_{-\infty}^{+\infty} x \times f(x)\,dx \int_{h}^{h+a_{SCALE}} = \int_{h}^{h+a_{SCALE}} x \frac{1}{a_{SCALE}}\,dx$$

$$= \left[\frac{x^2}{2a_{SCALE}}\right]_{h}^{h+a_{SCALE}} = \left((h + u_{SCALE})^2 - h^2\right)\frac{1}{2a_{SCALE}} \quad \bar{x} = h + \frac{a_{SCALE}}{2}$$

The mathematics has shown that the average reading is half a scale increment above the scale increment, which should not be a surprise.

The variance of the distribution calculated about the mean value is given by:

$$var = \int_{-\infty}^{+\infty} (x-\bar{x})^2 f(x)dx = \int_{h}^{h+a_{SCALE}} \left(x-h-\frac{a_{SCALE}}{2}\right)^2 \frac{1}{a_{SCALE}}dx$$

$$= \left[\frac{\left(x-h-\frac{a_{SCALE}}{2}\right)^3}{3a_{SCALE}}\right]_{h}^{h+a_{SCALE}} = \frac{a_{SCALE}^2}{12}$$

This variance is comparable to a Type A elemental uncertainty:

$$u_A = \sqrt{var} = \sqrt{\frac{a_{SCALE}^2}{12}} = \frac{a_{SCALE}}{\sqrt{12}}$$

For the previous examples that included rounding to the nearest scale increment, the average reading was the same as the true value. For this truncation example, the average reading is not the true value, and we have to include the systematic uncertainty this causes. The systematic offset is $a_{SCALE}/2$. Hence the standard uncertainty for truncation is given by:

$$u_{TRUNCATION} = \sqrt{(u_A)^2 + \left(\frac{a_{SCALE}}{2}\right)^2} = \sqrt{\frac{a_{SCALE}^2}{12} + \frac{a_{SCALE}^2}{4}} = \frac{a_{SCALE}}{\sqrt{3}}$$

An alternative derivation is to find the variance of readings from a scale increment rather than from the average value. Mathematically this is:

$$var = \int_{-\infty}^{+\infty} (x-h)^2 f(x)dx = \int_{h}^{h+a_{SCALE}} (x-h)^2 \frac{1}{a_{SCALE}}dx$$

$$= \left[\frac{(x-h)^3}{3a_{SCALE}}\right]_{h}^{h+a_{SCALE}} = \frac{a_{SCALE}^2}{3}$$

And the standard uncertainty is

$$u_{TRUNCATION} = \sqrt{var} = \sqrt{\frac{a_{SCALE}^2}{3}} = \frac{a_{SCALE}}{\sqrt{3}}$$

Appendix

It doesn't matter what mathematics you use, you get the same answer!

Thus, when using a digital display that truncates the reading, the standard uncertainty is:

$$u_{TRUNCATION} = \frac{a_{SCALE}}{\sqrt{3}}$$

REMEMBER, a_{SCALE} is the smallest reading (least significant digit) shown on the display.

ALSO NOTE that if the digital display rounds the number shown to the nearest division then the standard uncertainty is the same as for a rounded analog scale:

$$u_{ROUNDED} = \frac{a_{SCALE}}{\sqrt{12}}$$

Truncation with Error Adjustment

You realize that when a digital display truncates its readings, the average displayed value is half a scale increment lower than the true value. In this case, you could include this systematic uncertainty (or error), to every measurement. The uncertainty is then the same as when we have rounding.

For example, the temperature displayed on a digital display is truncated to 0.5°F. If you measure a temperature of 78.0°F, what is the standard uncertainty of this measurement due to scale resolution?

If you use the values of 78.0°F in your analysis, the standard uncertainty is given by the truncation equation:

$$u_{TRUNCATION} = \frac{a_{SCALE}}{\sqrt{3}} = \frac{0.5}{\sqrt{3}} = \pm 0.2887 F$$

However, adding the systematic uncertainty (half of the scale size) means that you would use the value 78.25 °F in your analysis, and the standard uncertainty due to resolution is given by the rounding equation:

$$u_{ROUNDED} = \frac{a_{SCALE}}{\sqrt{12}} = \frac{0.5}{\sqrt{12}} = \pm 0.1443 F$$

The Standard Uncertainty for an Analog-to-Digital Converter

Analog-to-digital (A/D) converters produce a stepped output for a smooth input. While instrumentation grade A/D have very small steps, they still introduce some uncertainty which is sometimes referred to as the "digitization noise". We can determine the standard uncertainty fairly quickly by recognizing that the individual steps in the A/D are equivalent to the scale increment for truncation, derived above. The A/D increment $a_{A/D}$ is related to the maximum converter input and number of bits in the A/D by:

$$a_{A/D} = \frac{V_{MAX}}{2^n}$$

Where V_{MAX} is the maximum voltage (or other engineering units if appropriate) that can be applied to the A/D without overload, and n is the number of bits in the A/D converter. Depending upon the way an A/D is implemented, the effective number of steps including V_{MAX} could be 2^n or $(2^n - 1)$. Rather than needing deep knowledge of the internal coding of your A/D, we counter with the argument that if the difference in measurement uncertainty between 2^n or $(2^n - 1)$ steps is significant to your measurement process, then you have already identified that the A/D is a problem and needs looking into!

We continue assuming there are 2^n steps in the A/D.

Many simple A/D converters truncate the values. Comparing the scale increment size for the A/D with the previous truncation derivation, we can quickly identify that the standard uncertainty due to digitization is given by:

$$\text{with truncation, } u_{A/D} = \frac{V_{MAX}}{2^n \sqrt{3}}$$

Some high-end converters use much more elegant data capture and signal processing algorithms that do not directly round or truncate[4], but can be treated as if they round. In this case, the standard uncertainty due to digitization is given by:

$$\text{with rounding, } u_{A/D} = \frac{V_{MAX}}{2^n \sqrt{12}}$$

---

[4] In an email to the authors from Oros, Inc, a manufacturer of high-end real-time multi-analyzers, they state that they extend the digital sample resolution from 24 to 32 bits floating words in order to avoid computation noise being visible in the results. They state this is closer to a rounding technique than a truncation, but the mathematics is not formal.

If you do not know whether you're A/D truncates or rounds, assume that it truncates. This will give a higher standard uncertainty. If, at the end of your uncertainty analysis, you identify that the A/D converter is the critical component then the A/D needs investigating.